PROFESSIONAL AND TECHNICAL WRITING STRATEGIES

PROFESSIONAL AND TECHNICAL WRITING STRATEGIES

Judith S. VanAlstyne
Broward Community College

PRENTICE-HALL, INC., ENGLEWOOD CLIFFS, NEW JERSEY 07632

Library of Congress Cataloging in Publication Data

VanAlstyne, Judith S. (date)
Professional and technical writing strategies.

Includes index.
1. English language—Rhetoric. 2. English language—Business English. 3. English language—Technical English. I. Title.
PE1479.B87V36 1986 808'.066 85-3510
ISBN 0-13-725813-5

Editorial/production supervision and
interior design: Chrys Chrzanowski and Virginia McCarthy
Cover design: 20/20 Services, Inc.
Manufacturing buyer: Harry P. Baisley

Printed in the United States of America

10 9 8 7 6 5 4 3

ISBN 0-13-725813-5 01

Prentice-Hall International, Inc., *London*
Prentice-Hall of Australia Pty. Limited, *Sydney*
Prentice-Hall Canada Inc., *Toronto*
Prentice-Hall Hispanoamericana, S.A., *Mexico*
Prentice-Hall of India Private Limited, *New Delhi*
Prentice-Hall of Japan, Inc., *Tokyo*
Prentice-Hall of Southeast Asia Pte. Ltd., *Singapore*
Editora Prentice-Hall do Brasil, Ltda., *Rio de Janeiro*
Whitehall Books Limited, *Wellington, New Zealand*

CONTENTS

PREFACE

Although this text is designed primarily for the two year college student seeking an Associate of Arts degree, it is comprehensive and flexible, suitable for college and university students at any level, professional and technical writers in the field, and business people seeking a guideline and model text. The materials have been tested in academic classes and in training workshops in a variety of businesses and industries.

Students of the 1980's are a heterogeneous group representing a myriad of interests and needs. They range in age from 17 to 70 (average age, 28) and major in everything from aviation to zoology. The text includes writing samples which illustrate actual writing demands in a cross section of career fields: allied health professions, criminal justice, data processing, electronic technology, fire science, landscape technology, pest control technology, insurance, and real estate, to name a few. These samples have been culled from students as well as from professionals in business and industry.

The emphasis is on practical writing and its applications, rather than on rhetoric and theory. The book covers strategies for writing effective correspondence, professional reports, and technical manual components. It also focuses on writing research and documented reports (employing both the new and traditional MLA style) and on building oral communication skills.

Although organization of the text intends to offer cumulative skills—moving from general considerations, to correspondence, to brief reports, to longer reports, and, finally, to specialized technical writing strategies—individual instructors may move about the text as freely as their audience and purpose dictate. Each chapter provides a list of skills which should be obtained, writing strategy guidelines, samples, exercises to reinforce the strategies, and writing options.

The instructor's manual offers general notes to the instructor, student preparation guidelines, and sample syllabi for a variety of class situations. In addition, the manual discusses approaches to each chapter, provides exercises answers, and illustrates solutions to the writing options.

ACKNOWLEDGMENTS

I wish to thank Irene VanAlstyne, Jeanne and Chuck Laird, Mary and Jerry Noosinow, Dr. Margaret Maney, and Chas Howard for their assistance and caring support. Laquita Withrow, Fran Brown, and Neil Linger offered invaluable suggestions, editing, research, and index assistance.

Thanks also to the staff at Prentice-Hall: Rolando Hernandez, Sales Representative; Phil Miller, English Editor; Chrys Chrzanowski, Virginia McCarthy, and Irene E. Springer, College Book Editorial/Production; Michael Tubridy, College Advertising, and all of the others who played a part in the production of this text. A sincere word of appreciation is due Joseph T. Barwick, Central Piedmont Community College; Lucas Carpenter, Suffolk County Community College; Warren Cushine, Hudson Valley Community College; Gloria Johnson, Broward Community College-North; Jack Klug, San Antonio College, and Louis E. Murphy, Bucks County Community College, for their rigorous reviews of the manuscript.

A special thanks to my contributors: students in both my academic classes and professional workshops, business and industrial professionals, and the companies who so willingly gave me permission to reproduce their correspondence, reports, and manual materials.

And a heartfelt thank-you to Kent and Jeff Duckham for their patience.

PROFESSIONAL AND TECHNICAL WRITING STRATEGIES

1

PROFESSIONAL/TECHNICAL COMMUNICATIONS

Skills:

After studying this chapter, you should be able to

1. Define *audience, purpose, style, tone, message, jargon, shop talk, gobbledygook, wordiness, redundancy,* and *voice.*
2. Recognize that the anticipated audience determines the manner in which the message should be written.
3. State the purpose of a message quickly and clearly.
4. Write brief messages in factual, objective, impersonal style.
5. Avoid an emotional, flowery, judgmental, or pompous tone.
6. Organize an attractive message by using short paragraphs, visuals, and graphics.
7. Recognize and eliminate jargon, shop talk, and gobbledygook.
8. Recognize and eliminate wordiness.
9. Recognize and eliminate redundancies.
10. Recognize and revise sluggish passive-voice constructions.

INTRODUCTION

Television, computers, movies, advertising flyers, sales promotion letters, highway signs, and shop marquees bombard our sense of sight and literate ability. Word processors and copy machines spew out language in mind-boggling quantity. The transactions of all occupations (business, industry, services) depend largely upon written correspondence and reports—concise, orderly messages which elicit results, not confusion.

Even the entry-level employee—the clerk, the fire fighter, the pest control technician, the physical therapist—must contribute to the written record of the company. Frequently, the employee who writes well is the one who is noticed by management and marked for promotion. Your ability to write reveals your organizational skills, your persuasiveness, and your logic. Your writing may even suggest your commitment to your occupation or profession.

Surveys show that employees at all levels spend approximately 25 percent of their time at work writing. Further, another significant percentage of time is spent dealing with the writing of others. And finally, despite the diversity of careers in the nation, writing demands are more similar than dissimilar. On-the-job writing includes a variety of special purpose letters, memoranda, long and short reports, sets of instructions, and explanations of procedures and mechanisms.

The occupational writer is faced with a number of questions:

- Who will read my message? (Audience)
- Exactly what is my message to accomplish? (Purpose)
- What language level should I use? (Style)
- What emphasis should I choose? (Tone)
- Just exactly what shall I write? (Message)

Sorting out the answers to these questions is the key to effective professional and technical communications.

AUDIENCE

"Different strokes for different folks" is a wise adage. Anticipating your audience is a major key to writing well. Your audience is the reader or readers to whom your written message is directed. You must ask yourself still more questions to determine audience:

- Is my audience informed or uninformed on the subject?
- Does my audience have a grasp of the specialized vocabulary of my field or not?
- Is my audience singular or a group?
- Is my audience a subordinate, a peer, or a superior in my occupation?

Just as you would make natural adjustments in vocabulary, sentence structure, and overall message in explaining the operation of a toaster to a child or to an adult, so will you make subtle changes in written instructions of any sort to a potential customer or to a group of factory assemblers.

If your reading audience has received preliminary information on a subject, you may eliminate details which are old information. If your audience is uninformed about the subject, you must be alert to include definitions, descriptions, background data, and other pertinent information.

Determine ahead of time if your reading audience has specialized training and vocabulary. If the audience's background is similar to your own, you may include technical terms, abbreviations, and graphics which an expert understands. If, however, your expected audience has little training in your field, you must remember to include synonyms, explanations, and details in plain English.

Some memorandums, letters, and reports are addressed to a single individual while others are intended for groups. The number of potential readers determines the style and tone in which the message is written. When you write for a single individual, you may be more personal, but when you write for a group, such as members of your department or all employees of an organization, or even potential readers of a manual or a professional journal article, a more objective and detailed report will be necessary. You need to identify the *lowest* level of understanding within the group and write to that level.

Whether your audience consists of subordinates, peers, or superiors should have a bearing on your presentation also. You must determine whether you should be familiar, supportive, or gracious.

Use a "you" perspective in all of your writing. A "you" perspective does not mean that you use the second person pronoun, but that you keep your audience's needs in mind throughout your correspondence or report. Put yourself in the reader's place and write naturally, concisely, and objectively. Audience consideration is so important that it is discussed separately in the following chapters for each type of writing strategy.

PURPOSE

By keeping one clear-cut purpose in mind, you will write a better letter or report. Purpose is your intention, aim, or plan. Is your purpose to query? to explain? to analyze? or to persuade? Not only should your purpose be clear to you, but you should also establish your purpose as quickly as possible in your writing. Begin your writing with such statements as:

> I am seeking additional information on your IBM Selectric II Typewriter.
>
> This pamphlet has been prepared to help you use your Ace microwave oven safely and efficiently.
>
> The feasibility of purchasing an Apple II computer for use in our three flower shops is presented in this study.
>
> The purpose of this proposal, regarding a salary index scale, is to provide incentive to our employees to remain with the company and to eliminate the charge of "favoritism" which has contributed to low employee morale.

Knowing and establishing your purpose will keep your message "on track" and will give your reader a clear sense of how to handle your information.

STYLE

Occupational writing usually uses an impersonal and simple style. Style is the manner or mode of expression in language, your way of putting thoughts into words. Unless your audience is technically sophisticated, you will want to use simple words, uncomplicated sentences, and short paragraphs. Avoid technical terms, jargon, shoptalk and gobbledygook, and overblown language. You will also want to avoid wordiness, redundancy, and unnecessary passive-voice constructions. Inappropriate style presents more problems in occupational writing than does any other consideration. If a reader cannot understand your language, your message is lost. This chapter will emphasize stylistic problems and how to overcome them.

TONE

Almost all occupational writing demands a factual and objective tone. Tone is the word choice and phrasing which expresses your attitude. Occupational writing is marked by its lack of emotionalism, editorializing, sarcasm, or even overt enthusiasm. Consider this partial text of an incident report illustrating inappropriate tone:

> On Wednesday, May 26, 198X, at 10:42 A.M., I had the misfortune of witnessing Keypunch Operator Polly Black take a nasty fall in the 5A West office area.
>
> While not looking where she was going, Polly clumsily caught her heel on a CRT tri-stand and crashed to the floor. The paramedic on call at security rendered first aid and transported her to Mercy Hospital. The accident was due to sheer carelessness.

The tone in this report is sarcastic (*had the misfortune*), judgmental (*while not looking where she was going, clumsily*), emotional (*a nasty fall, crashed*), and pompous (*CRT tri-stand, rendered, transported*). The following, a revised version of the same incident, illustrates an objective tone:

> On Wednesday, May 26, 198X, at 10:42 A.M., I witnessed Keypunch Operator Polly Black fall in the 5A West Office area.
>
> While approaching the door, Ms. Black caught her heel on an equipment stand and fell on her left side. The paramedic on call at Security treated her bruised knee and took her to Mercy Hospital.

The tone of occupational communication should be factual and impartial.

MESSAGE

Once you have identified your audience and purpose and you have considered your style and tone, you are ready to consider your message. Your message is the explanation, response, set of instructions, recommendations, or questions which will accomplish your transaction. But before you write, you should consider your organization, format, and other special features, such as visuals and graphics.

The reader probably has piles of paperwork to handle—letters requiring a response, memos giving instructions for action, bulletins and reports imparting information to be retained, and more. The last thing the beleaguered reader wants to read is a "grey" message, a page or pages of uninterrupted type. To avoid the "grey" look:

- Be as brief as possible.
- Divide your message into an introduction, body, and closing.
- Write in short paragraphs (4 to 5 typed lines).
- Use headings (Completed Work, Work In Progress, Work To Be Completed).
- Use numbers for key points or chronological steps (1., 2., 3.).
- Use capital letters for section headings (COSTS, PERSONNEL).
- Use underlining, v a r i a b l e s p a c i n g, and asterisks (*) to indicate divisions and headings.
- Use bullets (•) for lists of items.
- Insert graphics (bar charts, line graphs, drawings) such as those discussed in Chapter 4.

Although a message may be as brief as two sentences, it may also be many pages. Careful attention to visuals will make your message more readable.

CONCRETENESS

Guard against unfounded assumptions about your audience. DON'T! Even though you think your reader or readers *should* know why you are writing, *should* have a grasp of the situation, *should* be familiar with the language you are using, or *should* be able to act upon your message, you may be wrong. People are busy and preoccupied with their own work. Do not be vague. Use concrete language to eliminate any possible questions in the mind of the reader.

Vague	As we discussed recently, I have the figures on the project.
Concrete	I have the comparative costs of three word processing computers which you requested in our telephone conversation last Friday.
Vague	The policy change will affect us adversely.
Concrete	New Policy 1204.05 (Leaves) will decrease our allowable sick days from 10 to 8 per year.
Vague	We will fill your order within the next few weeks.
Concrete	We will ship C.O.D. your order for three, 4-drawer, 36 in. high by 45 in. deep by 14 in. wide, beige filing cabinets September 2, 198X. You should receive them no later than September 30.

JARGON, SHOP TALK, AND GOBBLEDYGOOK

Jargon is the specialized vocabulary and idiom of those in the same work. The jargon of one field often spreads to the professional world at large. Other words and phrases are intelligible only to those in the same line of work and may be classified as *shop talk*. Another type of jargon, which may be called *gobbledygook*, is characterized by unintelligible, pompous, or stiff language. General jargon and shop talk may be acceptable in very personal oral communication, but a competent writer eliminates all jargon from written communication.

General Jargon

The following list of words and phrases is used rather widely in informal occupational communications:

ballpark figure	interface
bottom line	optimization
finalized	output
game plan	parameters
impacted	time frame
input	viable

Moderate use of such terms is common in verbal transactions but should be revised for written messages. The following sentences demonstrate typical jargon exchanges and suggested written revisions:

Jargon	Give me a ballpark figure on the new office furniture.
Written	Give me a price estimate on the new office furniture.
Jargon	We'll use the input of each department to finalize our game plan.
Written	We will consider the suggestions of each department to complete our programming.
Jargon	The bottom line is that the recession has impacted on our hiring time frame.
Written	The key point is that the recession has affected our hiring schedule.
Jargon	The parameters for departmental interfacing must be viable.
Written	The guidelines for departmental boundaries must be realistic.

Shop Talk

The use of shop talk, the more technical slang of those in the same profession, becomes second nature to the users but should never be used in writing. Every occupation has its own shop talk.

Television shop talk	He shot the bridge with a minicam and then bumped up the tape.
Translation	He filmed the connecting segment between the news items with a small camera and then machine-processed the video-tape to a larger size.
Aviation shop talk	He checked the pax list, activated the SATCOM, and prepared the PIREP.
Translation	He checked the passenger list, turned on the satellite communication system, and prepared the pilot's report on meteorological conditions.
Academic shop talk	The increase in FTE's is probably due to so many students' having clepped math.
Translation	The increase in the number of full-time equivalency students is probably due to the many students who have waived the mathematics requirement by passing the College Level Examination Program test.

Gobbledygook

Besides avoiding jargon and shop talk, the skillful writer should avoid gobbledygook—unintelligible, pompous, and stiff language. Gobbledygook may sound more official or important but rarely states the message clearly.

Gobbledygook	At this juncture, the aforementioned procedure should be utilized.
Plain English	The plan which we discussed should be used now.
Gobbledygook	We should commence operational capabilities in systematic increments.
Plain English	We should begin the project step-by-step.
Gobbledygook	It would be prudent to consider expeditiously the provision of instrumentation that would provide an unambiguous indication of the level of fluid in the reactor vessel.
Plain English	We need a more accurate device to measure radioactivity.

WORDINESS

Effective professional writing is characterized by its brevity. The concise writer avoids roundabout phrases, redundancies, and sluggish passive-voice constructions.

Following is a checklist of shorter words and phrases to replace wordy, roundabout phrases:

Roundabout phrases	***Concise expressions***
a downward adjustment	cut, decrease
a great deal of	much
a majority of	most
accounted for the fact that	because
affix a signature to	sign
after the conclusion of	after
as a result	so, therefore
as a result of	because
as per your request	as you requested
as soon as	when
at which time	when
at all times	always
at an early date	soon
at a much greater rate than	faster

Roundabout phrases	*Concise expressions*
at the present time at this time	now
at the time of	during
avail yourself of	use
based on the fact that	because
be acquainted with	know
be of assistance to	assist, help
brief in duration	short, quick
by way of	by, to
came to an end	ended
consensus of opinion	opinion, everyone thinks
despite the fact that	although, though
due to the fact that in view of the fact that	because
enclosed please find	here is
for the purpose of	for, to
for this reason	so
for the reason that	since, because
give encouragement to	encourage
he was instrumental in	he helped
higher degree of	higher, more
in a manner similar to	like
in a position to	can
in accordance with	by, under
in favor of	for, to
in lieu of	instead
in reference to in relation to	on, about
in the amount of	of
in the nature of	like
in the vicinity	near, around
is dependent upon	depends on
is situated in	is in
it is necessary that	you must
it is recommended	we recommend
miss out on	miss
not infrequently	often
on account of	because, due to
on the part of	from, of

preparatory to prior to	before
provided that	if
pursuant to	under, with, following
referred to as	called
so as to	to
through the use of	by, with
to the extent that	as far as
until such time as	until
with reference to with regard to	on, about
with the exception of	except
with the result that	so that

Due to the fact that your reader has a great amount of other work to account for, it is necessary that you write so as to eliminate wordiness through the use of concise words. In other words, because your reader is busy, write concisely.

REDUNDANCY

A *redundancy* is a phrase which says the same thing twice (*repeat again*), contains obvious expansion (*square in shape*), or doubles the idea (*each and every*). Such repetition is pointless, wordy, and distracting. Consider the redundancies in the following sentences:

Redundant	It is *absolutely essential* in *this day and age* to *completely eliminate bigotry and prejudice.*
Revision	It is necessary now to eliminate bigotry.
Redundant	The sheriff's department and the city police *cooperated together* in *the month of May* to *devise and develop* a *totally unique drug and narcotics* control program.
Revision	The sheriff's department and the city police cooperated in May to develop a unique narcotics control program.
Redundant	There are *many in number* who consider the *total understanding* of *basic fundamentals* as a *good asset.*
Revision	Many consider the understanding of basics as an asset.

VOICE

Verbs have two voices: active and passive. In an active-voice expression the subject of the sentence performs the action stated by the verb.

Mr. Jones *conducts* the plant tours.
The president *presented* the budget.
The firm *is spending* $50 million this year to promote the new beer.
The new price *will increase* our profit.

In passive voice expressions the subject is acted upon.

The plant tours *are conducted* by Mr. Jones.
The budget *was presented* by the president.
Fifty million dollars *will be spent* this year by the firm to promote the new beer.
Our profit *will be increased* by the new prices.

Generally, the active voice suggests immediacy and emphasizes the subject. Because the passive-voice verb is always at least two words (the verb plus a form of *to be*), passive-voice expressions tend to be wordy. The passive voice may bury your main idea in sluggish sentences.

Edit your writing to determine which voice permits your desired emphasis in the fewest number of words.

The secretary typed the report. (Emphasis on secretary)
The report was typed by the secretary. (Emphasis on report)

EXERCISES

Eliminate *assumptions*. Remove all vagueness in the following sentences by inventing as many concrete details as necessary to answer any possible questions of the reader.

1. The work is now quite a bit behind schedule.
2. We need frequent inspections at critical checkpoints despite the cost suggested by your representative during her recent visit.
3. The typewriters in our department are frequently not as efficient as those in yours.
4. We will implement your proposal provided that the requisite labor can be found.
5. Those letters should be attended to properly.

Eliminate *jargon and gobbledygook*. Rewrite these sentences into simply expressed, intelligible sentences.

1. At this juncture, the bottom line depends on sales output.
2. In a nutshell we failed to finalize our conceptualized contingencies in the proper time frame.
3. We require in-depth communication to determine why our game plan was defunded.
4. During the implementation phase the recession impacted upon our employment levels.
5. Research indicates that the distinction of critical sociocultural parameters rendered the project nonviable.

Eliminate *wordiness*. Rewrite these sentences to eliminate wordy and roundabout phrases.

1. We are inclined to make the recommendation to utilize the room for the purpose of training.
2. Due to the fact that the seat belt broke, the passenger sustained a high degree of injury.
3. Pursuant to our discussion, we decided to let go some forty technicians until such time as the economy picks up.
4. If we plan on showing a profit of three percent, we will need to make a downward adjustment in travel expenditure.
5. As per your request, I have affixed my signature to the reports at this time.

Eliminate *redundancies*. Rewrite these sentences to eliminate repetitions, unnecessary expansions, and doublets.

1. Either one or the other of the copy machines is totally acceptable.
2. It is absolutely essential that each and every secretary employ the use of new typewriter ribbons.
3. The troops advanced forward in close proximity to the enemy territory.
4. Through mutual cooperation we can stamp out and eliminate crime in this day and age.
5. During the month of May we will begin to package our product in boxes square in shape and red in color.

Eliminate sluggish *passive-voice* expressions. Rewrite these passive constructions into the active voice.

1. A report on the salary increases was requested by the union officers.
2. Graphic construction is discussed in the next chapter.

3. For the final report a cover letter was used.
4. At our corporate headquarters decisions are made.
5. Jargon is in the chapter in which style and tone are discussed.

WRITING OPTIONS

1. *Language.* Rewrite the following memorandum in an appropriate style and tone. Divide the message into brief, logical paragraphs. Eliminate jargon, gobbledygook, wordiness, redundancies, and awkward passive voice expressions.

   MEMORANDUM

TO:	Nancy Hensley, Training Specialist
FROM:	Ralph Singlefoot, Secretary
SUBJECT:	Coffee Committee
DATE:	January 7, 198X

   As per your request, I am accordingly submitting to you a report on the Committee for the Coffee Project. Several meetings have been held which were attended by Ms. Sandra Harper, Mr. Lincoln Jones, and me to look into the matter of the workroom coffee pot situation. Ever since the Proctor-Silex 30-cup coffee pot was broken, we have been totally without coffee. Furthermore, each and every one of the cups which was available for use has disappeared. This is a disgrace. Perhaps some kind soul has a suggestion as to how to handle this. At this point in time we are endeavoring to ascertain the total sum of money we will absolutely require in order that we can expedite the coffee delivery system for the betterment of the department. A ballpark figure for the project is $100.00. It is the consensus of opinion that a new pot and cups are needed and should be rapidly purchased to begin the program quickly. We would like to recommend that this problem be placed on the agenda listing for discussion at the next Tuesday departmental meeting in the month of May at 3:00 P.M. in the afternoon. If you have any input of ideas we would assuredly like them before the next meeting.

2. *Message.* Reorganize the following report to make the message more readable. Divide the material into logical paragraphs; add section headings, numbers, bullets, and any other appropriate visual device to make the proposal more readable.

   Here is a proposal to introduce a standard format for sales proposals in order to avoid inconsistent layout, to improve secretarial productivity, and to save the company money. Because the sales force has been relocated from headquarters into the field, the system of preparing the sales proposals has presented the following problems: salespersons must dictate all proposals by costly telephone calls, each dictation suggests a different format which is confusing, the salespersons require a mailed copy for approval and editing; each then mails the proposal back with format changes, an inefficient procedure. To cut costs, eliminate inconsistent proposal formats, and to improve efficiency, I propose that we should develop a form for each salesperson to fill in with the particular prices, routing, and dates for each sales proposal. Secondly, we should create a printed sample proposal format for each salesperson to use consistently. The information sheet may be mailed or phoned to the office, and inserted into the standard format for immediate mailing to the client. Attached are the two sample forms. The advantages of this system are long distance telephone calls will be eliminated or shortened, sales persons will have a checklist which will eliminate mistakes and omissions of details, all proposals will be consistent, and the preparation of the sales proposals will be faster and more efficient. I am available to discuss this proposal with you Monday through Thursday.

3. *Audience.* Rewrite a short (3–4 paragraph) article from your field. Select an article or textbook explanation which is as technical and complex as you can understand and then translate the material for a different audience, such as a junior high school student or a layperson who has no knowledge of your field. Submit a copy of the original along with your rewrite.

2

CORRESPONDENCE

Skills:

After studying this chapter, you should be able to

1. Recognize conventional memorandum formats.
2. Write a brief, focused memorandum in an appropriate format.
3. Lay out letters in full block and modified block formats.
4. List, define, and write an example of the regular and special elements of letters.
5. Head a second page of a letter correctly.
6. Write an effective "good news" letter: an inquiry or request, an order, a congratulatory or thank-you letter, or a sales or service offer.
7. Write an effective "bad news" letter: a negative response, a complaint, a collection, or a solicitation letter.
8. Prepare a business letter envelope correctly.

INTRODUCTION

The writing strategies professional persons need first in all career fields are those required for effective correspondence. Memoranda and letters are records which promote action, transact business, and maintain continuity.

A one-page typewritten letter can cost up to $7.50 in terms of equipment, salary, paper, and postage. If correspondence is being generated by a word processing computer, the cost is even higher. Therefore, mastering the strategies of effective correspondence is crucial for cost-effective business transactions.

A memorandum, usually called a *memo*, is used for internal (within the company or organization) correspondence. A memo is used to request information, confirm a conversation, announce changes in policy and procedure, congratulate someone, summarize meetings, transmit documents, and report on day-to-day activities.

Letters are used for external (outside of the company or organization) correspondence. A letter may be used to seek employment (see Chapter 3), request information, respond to inquiries, promote sales and services, register complaints, offer adjustments, provide instructions, or report on any other activity.

Both types of correspondence are characterized by a "you" perspective—brevity, clarity, and accuracy.

MEMORANDA

Types of memos may vary from a brief, handwritten message which will probably be discarded by the reader after the appropriate action is taken to a carefully typed message which will be filed as part of the permanent record.

Format

Many companies use a preprinted communications memorandum for telephone and caller messages as shown in Figure 2.1:

To ______________________

Date ______________ Time __________

WHILE YOU WERE OUT

M ______________________

of ______________________

Phone ______________________

Area Code Number Extension

TELEPHONED		PLEASE CALL	
CALLED TO SEE YOU		WILL CALL AGAIN	
WANTS TO SEE YOU		URGENT	
	RETURNED YOUR CALL		

Message ______________________

____

Operator

Campbell 09301

Figure 2.1 Preprinted communication memorandum

Some organizations also use preprinted memo forms on 8½ by 11-inch or half-page paper. These usually include the company name and logo plus the guide words *TO, FROM, SUBJECT,* and *DATE*. Some organizations use a "speed memo," one with carbon copies which allow the reader to send the message, retain a copy, and solicit a reply, as shown in Figure 2.2:

R RYDER **RYDER TRUCK RENTAL, INC.** Date ____________

TO ▸ ____________ At ________ SUBJECT: ____________

FROM ____________ At ________ LOCATION CODE: ____________

MESSAGE ____________

____________ Signed ____________

Date ____________

REPLY ____________

____________ Signed ____________

S10-01 (1-79)

ADDRESSEE TO RETAIN WHITE COPY — RETURN GREEN COPY TO ORIGINATOR
ORIGINATOR TO RETAIN BLUE COPY — SEND WHITE AND GREEN COPIES WITH CARBON INTACT TO ADDRESSEE

Figure 2.2 Preprinted speed memorandum *. By permission.*

If your company does not have preprinted forms, the model shown in Figure 2.3 is the most common format:

MEMORANDUM
(2 lines)
TO: John Snyder, President
(2 lines)
FROM: Maureen O'Neill, Personnel Specialist M.O.
(2 lines)
SUBJECT: Fire in Reception Area, 1/13/8x
(2 lines)
DATE: January 14, 198x
(2 lines)
____
(2 lines)
____
____
________________________.
(2 lines)
____
____
________________________.
(2 lines)
MO:jv
(2 lines)
Enclosures

Figure 2.3 Standard memorandum format

Notice the capitalization, margins, and spacing. The message paragraphs are usually single spaced with double spacing between paragraphs. The writer's signature or initials appear after the typed name in the *FROM* line for authorization, and for possible legal purposes, too.

Although frequently written in haste, a memo requires careful consideration. The addressee line should contain the full name and title of the person. Most employees wear several hats within an organization. An employee may be the Director of Human Resources, Chairman of the Comptroller Search Committee, and member of the Ad Hoc Committee on Leave Policies. The receiver of a memo wants to know immediately in which capacity he or she is being addressed. Memos may also be addressed to several people.

Examples

TO: John Q. Jones, Secretary, Vacation Committee

TO: Dr. Glen R. Rose, Academic Dean
Dr. Benjamin Popper, Acting Director, Communications Division
Dr. Mary Ellen Grasso, Chair, English Department

The sender should also include one's own title, indicating the position from which the memo is written.

Examples

FROM: Susan Long, Professor of English

FROM: Susan Long, Chair, Ways and Means Committee

The subject line should contain a precise, descriptive title indicating the topic of the memorandum

Imprecise SUBJECT: Meeting report
Precise SUBJECT: Ways and Means Committee Minutes—3/5/8X

Content

Other conventions govern the content. A memo is usually **brief**, one-half to one page long. Most employees receive numerous memos each day and must react to each message. Reading and reacting take time.

Memos get **to the point** quickly. Usually the in-house reader is somewhat informed about the general subject. Lengthy background and tactful motivations to act are usually not necessary. Just state your message and exclude irrelevant detail.

Even though you have several messages to convey to the same person, do not confuse your reader. Cover only **one topic**, and write a separate memo for each new subject.

Finally, be **timely**. Announce meetings several days in advance. Respond in writing to the memos you receive as quickly as possible. Interoffice correspondence is designed for efficiency. Figure 2.4 shows a sample of a memorandum which is brief, to-the-point, and timely on a single subject:

MEMORANDUM

TO: Members of the Long-Range Planning Committee

FROM: Kenton Duckham, Chairperson K.D.

SUBJECT: Change in May meeting date

DATE: April 27, 198X

Because President Jean Laird must attend a marketing conference in Atlanta, the regularly scheduled May 15, 198X meeting of the Long-Range Planning Committee is rescheduled for May 22, 198X, at 3:00 P.M. in the Board Room.

Please notify me at extension 6647 if you will be unable to attend.

KD:jv

Figure 2.4 Sample memorandum

LETTERS

Effective letter writing is the key to successful business operation. Letters are the essential link between you and your business connections and between one organization and another. Letter content promotes and confirms most business transactions. Conventions of format, mechanics, and content regularize and simplify these transactions.

Format

Several acceptable letter formats are in use today. Individual companies tend to adopt a uniform format for use on preprinted letterhead stationery.

If letterhead stationery is not used or if you are writing business letters as an individual, select white, unlined, 8½ by 11-inch bond paper of about 20-lb weight. Cheap, flimsy paper neither feels nor looks important. The quality of the paper may seem like a small point, but it does make a difference in the amount of attention your letter will receive.

Figures 2.5 and 2.6 show the two most widely adopted letter formats: **the full block** and **the modified block**. Study the spacing and indentation. Side margins are 1 to 1½ inches. The top margin is a minimum 1 inch but may be deeper if the letter is short. A letter usually has more white space at the bottom than at the top, and it is conventional to type the body, or text, of the letter over the centerfold of the page.

Parts of Letters

All letters have six major parts: the heading, the inside address, the salutation, the body, the complimentary closing, and the signature. In addition, many letters contain subject, reference, or attention lines; typist's initials; enclosure and distribution notations.

Headings. If you are using printed letterhead stationery, add only the date two spaces below the letterhead. If you are using plain, white stationery, type your street address, city, state, zip code, and the date.

Example

1414 Southwest Ninth Street
Fort Lauderdale, Florida 33312
November 12, 198X

Do not include your name. Each line begins at the same margin. Refrain from abbreviating. Write out *Street, Avenue, Boulevard, East, Northeast*, and so on. Notice also that in street addresses one- and two-digit numbers are written out, but three or more digits are written in Arabic numerals.

Examples

One Landmark Plaza
Twenty-two West Third Avenue
123 Forty-second Street
but
2134 West 114th Street

Leave two spaces between the state and the zip code. The heading is placed flush left in the block format or flush to the right margin in modified block format.

Inside Address. The inside address includes the full name, position, company, and address of the recipient of your letter. It is spaced 3 to 6 lines below the heading.

Heading	409 Northeast Twelfth Avenue Fort Lauderdale, Florida 33315 February 12, 198X (3—6 lines)
Inside Address	Mr. James W. Nelson General Manager Ace Equipment Company, Inc. 1092 East Eleventh Street Seattle, Washington 98122 (2 lines)
Special Element	SUBJECT: Your letter of January 15, 198X (2 lines)
Salutation	Dear Mr. Nelson: (2 lines)
Body	______________________________ ______________________________ ____________________ . (2 lines) ______________________________ ______________________________ ______________________________ ______________________________ ____________________ . (2 lines) ______________________________ ______________________________ ______________________________ ______________________________ ____________________ . (2 lines)

Figure 2.5 Full block format for business correspondence

____

____

________________________ .

Compli-mentary Closing

(2 lines)
Very truly yours,

(4 lines)

Signature

Carol Hopper
(2 lines)
CH:cm

Notations

(2 lines)
c: David L. Meeks

Figure 2.5 Continued

Heading

2107 West New Boulevard
Fort Lauderdale, FL 33301
August 3, 198X

(3–6 lines)

Inside Address

Mr. Ray Adams, Associate
Sun Air Corporation
State Road 7
Fort Lauderdale, FL 33314

(2 lines)

Salutation

Dear Mr. Adams:

(2 lines)

Body

____
____
__________________________.

(2 lines)

____
____
____
____
__________________________.

(2 lines)

____
____
____
____
__________________________.

(2 lines)

Figure 2.6 Modified block format for business correspondence

____

____

__________________________________ .

(2 lines)

Compli-mentary Closing

Sincerely,

(4 lines)

Signature

Marcia Jordan
Personnel Specialist

(2 lines)

Notations

MJ:ts
(2 lines)
Enclosures (2)

Figure 2.6 Continued

Examples

Dr. Mary Jones, President
New Community College
101 South Palm Avenue
Miami, Florida 30301

Mr. Horacio L. Fernandez
Director of Human Resources
Ace Manufacturing Company, Inc.
Davenport, Iowa 90521

Notice that short titles may be placed on the same line as the name while long titles are placed in a separate, second line. If possible, address your letter to a specific person rather than just to a position within a company. While you may abbreviate titles such as *Dr., Mrs., Mr.,* and *Ms,* do not abbreviate *the Reverend* or *the Honorable* nor titles denoting rank, such as *Lieutenant, Captain, Professor.*

The inside address is always placed flush to the left margin.

You may abbreviate the state using the Postal Service two-letter abbreviations, as shown on page 29.

Salutation. The salutation is your greeting to your reader. It is typed two lines below the inside address and is followed by a colon. Further, it must agree with the addressee of the inside address.

Examples

Dr. Susan Clark, Dean
New Community College
101 South Palm Avenue
Miami, Florida 30212

Dear Dr. Clark: (agreement with person)

Director of Personnel
Ace Manufacturing Company, Inc.
4012 West Grand Street
Davenport, IA 90051

Dear Sir: (agreement with title)

Ace Manufacturing Company, Inc.
4012 West Grand Street
Davenport, IA 90051

Gentlemen: (agreement with corporate body)

Alabama	AL	Montana	MT
Alaska	AK	Nebraska	NE
Arizona	AZ	Nevada	NV
Arkansas	AR	New Hampshire	NH
California	CA	New Jersey	NJ
Colorado	CO	New Mexico	NM
Connecticut	CT	New York	NY
Delaware	DE	North Carolina	NC
District of Columbia	DC	North Dakota	ND
Florida	FL	Ohio	OH
Georgia	GA	Oklahoma	OK
Guam	GU	Oregon	OR
Hawaii	HI	Pennsylvania	PA
Idaho	ID	Puerto Rico	PR
Illinois	IL	Rhode Island	RI
Indiana	IN	South Carolina	SC
Iowa	IA	South Dakota	SD
Kansas	KS	Tennessee	TN
Kentucky	KY	Texas	TX
Louisiana	LA	Utah	UT
Maine	ME	Vermont	VT
Maryland	MD	Virginia	VA
Massachusetts	MA	Virgin Islands	VI
Michigan	MI	Washington	WA
Minnesota	MN	West Virginia	WV
Mississippi	MS	Wisconsin	WI
Missouri	MO	Wyoming	WY

League of Women Voters
Ten Northeast Datepalm Drive
Fort Lauderdale, FL 33302

Ladies: (agreement with gender)

Because many women are joining the corporate ranks of business and industry, it is not unusual to see salutations, such as *Ladies and Gentlemen:*, *Hello:*, or *Dear Director of Training:*. Unless a woman has expressed a desire for *Miss* or *Mrs.*, use *Ms* (optional period) whether she is married or unmarried.

Avoid *Sir*: (too formal), *My dear Sir*: (too pretentious), *Dear Sirs*: (use *Gentlemen*:), and *To Whom It May Concern*: (too impersonal).

The salutation is always typed flush to the left margin.

Body. The body, the text of your letter, begins two lines below the salutation. Notice in Figures 2.5 and 2.6 that paragraphs are not indented in the block format but are indented five spaces in the modified block. Single space within paragraphs and double space between them. The body should fall over the center of the page.

Complimentary Closing. The complimentary closing is two lines below the body and is followed by a comma. Only the first word is capitalized.

Examples

Very truly yours,	(formal)
Yours truly,	(less formal)
Sincerely,	(emotional)
Respectfully,	(if addressee outranks you)
Cordially,	(warm)

The complimentary closing is flush to the left margin in block format and at the horizontal midpoint or at the heading margin in modified block format.

Signature. Your full name is typed four lines under the complimentary closing. Sign your name in black ink between the two. If the addressee is known to you, you may sign less formally than the typed signature.

Example

Very truly yours,

Judy Van Alstyne

Judith S. VanAlstyne

Special Elements. Occasionally, a subject, reference, or attention line is placed between the inside address and the salutation to alert the reader to the subject, file reference, previous correspondence, account number, or other emphasis.

Examples

Subject: Invoice #20947

ATTENTION: Mr. D. W. Clark

RE: Your letter of June 12, 198X

Such special elements are typed flush to the left margin two lines below the inside address but above the salutation.

Typist's Initials. If your letter is typed by someone other than you, place your initials in capital letters and the typist's initials in lower case letters flush to the left margin two lines below the typed signature. Use a colon or a virgule between them.

Examples

JSV:mt
or
JSV/mt

Enclosure Notations. If you send materials or documents with your letter, add an enclosure notation two lines below the typist's initials.

Examples

Enclosure
or
Enclosures (2)
or
Enc: Photocopy of Check #1029

Distribution Notation. If you are sending copies of your letter to other readers, add a distribution notation two lines below the last element.

Examples

cc: Mr. David Little, Chairman
c: Ms Patricia Hansen, President

Since most copies are now photocopies rather than carbon copies, the *c* is more conventional.

Second Pages

If a letter requires a second page, type the recipient's name, the page number, and the date in a block flush to the left margin or across the page.

Examples

Ms Sally Queen
Page 2
July 15, 198X
or
Ms Sally Queen -2- July 15, 198X

Dr. Richard Grande
1022 Northeast First Street
Fort Lauderdale, FL 33314

Ms Julie Maney
Director of Personnel
Ace Manufacturing Company, Inc.
4092 West Grant Street
Davenport, IA 90521

ENVELOPES

Your envelope should be 9½ by 4½ inches and of the same quality as your stationery. The recipient's name, title, company, address, city, state, and zip code are centered horizontally and vertically. As a guideline, begin 12 lines down from the top. Single space between lines. Your own name, address, city, state, and zip code are placed in the upper-left-hand corner. Use the Postal Service abbreviations for states.

CONTENT

Organization

Organize your message into three parts:

1. A brief introduction which states the purpose of the letter immediately unless you are conveying "bad news."
2. One or more body paragraphs which contain specific detail.
3. A conclusion which establishes goodwill or encourages your reader to act.

Introductory Purpose

State your purpose immediately. Do not just fill space until you get around to your purpose.

Vague Purpose	It became apparent about five years ago that computerized bookkeeping was to be the answer to the problems which were plaguing our bookkeeping department. Therefore, we would like to investigate your software program. . .
Clear Purpose	I am seeking answers to three questions regarding your software program, Computerized Automotive Reporting Service.

Other clear introductory purpose statements follow.

Examples

Here are the instructions for assembling the Ace 1 Trampoline which you requested by phone on September 20.

Please consider my resume and application for a junior management position at your Pompano resort.

This is in answer to your inquiry about leasing our trucks.

Your vacuum cleaner is repaired and is ready to pick up.

Congratulations on your promotion to Director of Affirmative Action.

You are right. You paid your bill exactly when you said you did.

"You"-perspective Body

Provide the details of your correspondence in the body, using a "you" perspective. Put yourself in your recipient's shoes and consider how that person will respond to your words. Be courteous, direct, and confident. Avoid a slangy, abrasive, pompous, or abrupt tone. Consider the abrasiveness of the following phrases:

Examples

Your department should shape up . . .

I demand that . . .

I am appalled at your slow response . . .

I beg to advise you of my intent to . . .

Rush me information on . . .

Much of our correspondence is highly repetitive ("Thank you for your order of July 25." "Our records indicate that your payment is past due."), yet trite and cliché expressions should be avoided.

Cliché Expressions	***Plain English***
Having received your letter, we . . .	We received your letter . . .
Pursuant to your request . . .	As you requested . . .

Cliché Expressions	*Plain English*
Per your memorandum . . .	As you noted . . .
Enclosed please find my report . . .	Here is my report . . .
It is imperative that you write at once . . .	Please write at once . . .
I am cognizant of the fact that my report is tardy . . .	I know that my report is late . . .
At the earliest possible date . . .	As early as possible . . .
I beg to differ with your . . .	I disagree with your . . .
Please be advised that the new policy . . .	The new policy is . . .
I hereby request that . . .	Please consider . . .
I beg to acknowledge receipt of your check . . .	I received your check . . .
We are in hopes that you succeed . . .	Good luck . . .

Every letter you write should sound fresh and conversational.

Purposeful Conclusions

Your conclusion provides an opportunity for you to urge action or establish goodwill. A brief closing should motivate the recipient to follow up on your letter or, at least, to feel favorable to you, as illustrated in these examples.

Examples

May I have your answer by March 12?

If you will call us within the next few days, we can send our sales representative to demonstrate our software capabilities.

I would like an interview and am available weekdays for the rest of this month.

Thank you for pointing out this problem.

I appreciate your services . . .

Do not, however, be obvious or presumptuous. Avoid "I want to thank you in advance for . . ." and "Please feel free to call me if you have further questions." An advance thank-you implies that you may be too lazy to write a proper thank-you when your request is fulfilled. The second closing is unnecessary; the recipient will call you if he or she has questions whether you invite a call or not.

Word Processing

A final word about the "you" perspective should be added. Word processing by computers has removed the drudgery from written correspondence by allowing organizations to create standard text for repetitive form letters. If you are already using a word-processing computer program, you know that you can delete anything from a character to a paragraph or relocate or insert words, phrases, sentences, and entire paragraphs with ease.

The word-processing capability, however, may tend to depersonalize our letters. Great care must be taken to maintain a friendly, "you"-oriented tone.

GOOD NEWS LETTERS

Recipients of many types of letters are happy to receive them. Letters may inquire about merchandise and services to which the recipient is pleased to respond. Other letters offer services or sales, place orders, tender a congratulation or thank-you, or transmit desired information. You will not only receive such letters but will also be called upon to write them in your career field. "Good news" letters include:

- Inquiry and Request letters
- Order letters
- Congratulatory letters
- Thank-you letters
- Sales- or service-offer letters
- Employment application and cover letters
- Transmittal letters

Inquiry and Request Letters

An inquiry or request letter is a "good news" letter in that the recipient stands to benefit from the writer's interest. Nevertheless, an inquiry or request solicits a response which will ask of the recipient time and, perhaps, effort. Therefore, certain strategies will help to motivate the recipient to respond quickly and accurately.

First, your introductory purpose statement may include the suggestion that you need an immediate response.

Example

I am seeking additional information on your TP-I Daisy Wheel printer because I plan to purchase one this month.

Secondly, you may clarify why you need this particular information.

Example

Because I edit our Company's in-house newsletter, it is essential that I purchase a printer which performs proportional spacing.

Third, you may subtly compliment the person or the company.

Example

I am interested in obtaining some additional details about your VID-80 Model III. Your company was the first to advertise such a modification more than six months ago, so I feel that you have the most expertise in this field.

Fourth, if you are asking questions, simplify and separate them into numbered items, make each specific, arrange them in a numerical table, and allow sufficient space between each to allow a jot of answers right on your letter.

Example

I have three questions which will affect whether or not I upgrade my system at this time:

1. What type of format does your CP/M use? Is it compatible with the system Radio Shack computers use?

2. Do schematic diagrams come with the unit? If not, are they available for an extra charge?

3. If the added memory is purchased, is there any way to use the additional memory when operating with TRSDOS or similar operating systems?

You may also consider enclosing a self-addressed, stamped envelope. This double strategy of leaving space to answer on your letter and sending a return envelope will allow the recipient to respond right on your letter and place your response in the outgoing mail immediately.

Finally, if it is appropriate, you may offer to share the results of your inquiry.

Example

Because I am compiling information on the technical writing programs of all Florida community colleges, I will be happy to share with you my final report.

Figure 2.7 shows a poor request letter. It exemplifies a "me" rather than a "you" perspective and is poorly organized and rude. Figure 2.8, however, illustrates a well-written request letter which motivates a quick response.

Order Letters

Another "good news" letter is the order letter, one that informs a seller that you want to purchase a product or service. Three writing strategies will help you to be clear and accurate about your specific order, shipping instructions, and method of payment.

First, accurately describe the product or service by specifying the name, brand, model, stock number, quantity, color, dimensions, unit price, and so on. Include an informal table for multiple product or service orders for easy reader reference.

Example

Please send me the following accessories listed in your SFA August Folio Collection:

198	3	bangles	blue	@$3.00 ea	$ 9.00
200	1	bead necklace	bl/bk	@$15.00 ea	15.00
333	2	simulated pearls	white	@$40.00 ea	80.00
				Total	$104.00
				Shipping	4.00
				Total	$108.00

Second, include your shipping instructions, such as first-class or third-class mail, Federal Express, railway express, or special mailing address, department or special attention notation.

Example

Mail these first class to the following vacation address which differs from my billing address:

Miss Jane Hansen
702 Ocean Shore Lane
Ogonquit, ME 90437

3072 Southwest Sixth Avenue
Fort Lauderdale, FL 33321
May 4, 198X

Mr. Fred Mandel
Radio-Electronics
200 Park Avenue
New York, NY 70908

Dear Mr. Mandel:

Demanding and Insulting

Please rush me information on how to convert my transistor output voltage to 6VDC usable voltage for my portable radio. I read your article in Radio-Electronics, but it confused me.

Unclear questions

What I need to know is can I use the same 300 Ohm pot or something else for my 9VAC. Can I use the same number rectifier for my radio? Can you recommend a zenior diode? What else do I need to know?

Rude

Thank you in advance for answering my questions.

Very truly yours,

Walt

Walter Matthews

WM:jsv

Figure 2.7 A poor request letter

7661 Hood Street
Hollywood, FL 33024
June 29, 198X

Mr. Jerry Diener
Vice-president
Smith-Corona Company
65 Locust Lane
New Canaan, CT 06840

Dear Mr. Diener:

Compli-mentary
To-the-point

I saw your excellent advertisement for a TP-I Daisy Wheel printer in the July 1983 issue of 80 Micro on page 95, and I am writing to you to obtain answers to five questions concerning this printer.

Clearly organized questions

1. How many characters per second does this printer print?
2. Can the TP-I printer perform proportional spacing?
3. Can the printer produce carbon copies? If so, how many copies can it make?
4. Is there an additional charge for the parallel data interface? Amount?
5. Are the printer control codes the same as those used on the Diablo 620?

Enabling

For quick response, you may jot the answers to my questions on this letter. Please also send me any brochures available on your TP-I printer.

Figure 2.8 ***An effective request letter (Courtesy of student Sandra J. Burnette)***

Urge to action with good reason

I would appreciate this information as quickly as possible so that I may take advantage of your rebate offer which ends on July 31. I have enclosed a stamped, self-addressed envelope for your convenience.

Very truly yours,

Sandra J. Burnette

Sandra J. Burnette

SJB:jsv

Enclosure

Figure 2.8 Continued

Third, mention the date needed if this is an issue, and, finally, specify your method of payment: enclosed check or money order, credit card charge number, C.O.D., installments, and so on.

Example

If this order cannot be filled in the next two weeks, please call me collect to arrange alternate shipping. Please charge the amount plus shipping to my credit card, SFA 1002-0009-3047.

Figure 2.9 illustrates a clear and accurate order letter.

Congratulatory and Thank-you Letters

Both of these "good news" letters are characterized by informality and friendliness. The salutation might address the recipient by a first name. The introduction should mention specifically the occasion for congratulations or appreciation. Add detail to underscore your sincerity and end with a warm complimentary close.

Figures 2.10 and 2.11 illustrate typical congratulatory and thank-you letters.

Sales and Service Offer Letters

Even though a sales or service offer letter is written to persuade the recipient to purchase a product or service, it may be considered a "good news" letter in that it offers to enhance the recipient in some manner. Five writing strategies can aid you in obtaining a favorable response.

First, identify and limit your audience. Determine the needs of this group and bear in mind exactly what you want your audience to do after reading your letter. The "you" perspective is critical for your desired response. Keep in mind what you can do for your reader throughout the letter.

Second, begin your letter with an attention-getting statement. You may ask a question, offer a free gift, employ a "how to" statement, or use flattery. In short, hook your reader into reading further.

Examples

You will receive a free vacation for two, a new car, a dream cottage, or other valuable gift simply by making an appointment to inspect our resort.

Here's how to save $100.00 on your next automobile purchase.

You made a smart choice by enrolling at Broward Community College. Now let us help you make a smart choice in selecting your college wardrobe.

Specific order

Mailing instructions

Payment details

Sunshine Aviation School
1701 Rio Vista Boulevard
Fort Lauderdale, Florida 33316
July 30, 198X

Sporty's Pilot Shop
Clermont County Airport
Batavia, Ohio 45103

ATTN: Aviation Department:

Gentlemen:

Please send us by first-class mail the following items which are listed in your 198X catalogue:

3128	A	Jumbo Computer	10@$7.95	$79.50
2071	A	Topcomp Runway Computer	10@11.95	119.50
2241	B	Sectional Timescale	10@18.50	185.00
				$384.00

Mark the package "ATTN: Instructor Jerry Nordstrom." We will expect this order by next Friday. If there will be a delay, please call me collect (305-491-7702).

I am enclosing a check for one-half of the order ($192.00) and will pay the balance on receipt of the materials.

Very truly yours,

Jerry Nordstrom

Jerry Nordstrom
Chief Flight Instructor

JN:dt

Enclosure

Figure 2.9 An effective order letter

ACE COMPANY, INCORPORATED
2900 Northeast Seventy-ninth Street
Fort Lauderdale, Florida 33301

January 5, 198X

Ms Paula Watkins
Tech Laboratories, Inc.
P.O. Box 37021
Miami, Florida 30321

Informal salutation

Dear Paula:

Source of knowledge

Specific message

The good news that you were promoted to Director of Personnel was in the Business News section of this morning's Miami Herald. Tech Labs certainly picked the right person for the job. Congratulations!

Underscore of sincerity

You've earned this advancement, Paula. Your consistent good work and extra efforts, such as your participation in our Broward County Career Festival, set an example for all of us in the field.

Friendly closing

Warmest wishes,

Ben

Figure 2.10 An effective congratulatory letter

3502 Southwest Palm Avenue
Fort Lauderdale, Florida 33314
March 7, 198X

Professor J. John Jenks
Community Service Division
Broward Community College
1 East Las Olas Boulevard
Fort Lauderdale, Florida 33301

Dear Professor Jenks:

To-the-point Informal

Thank you for forwarding your generous recommendation of me to Tech Laboratories, Inc. I got the job!

Specific details of appreciation

The word processing skills you taught me plus your pep talks on organization have opened new doors of self-confidence for me. I'm truly grateful for your interest.

I'll keep you posted on my advancement.

Friendly closing

Thanks again,

Ralph

Ralph Kennery

Figure 2.11 An effective thank-you letter

Third, call attention to the product or service's appeal. Persuade the reader that your offer is so desirable that he or she can't resist it.

Example

> Our time-share condominiums are caressed by gentle ocean breezes and within steps of your very own tennis courts, golf course, and spa. We offer the last word in glamorous vacations.
>
> Ace Motors offers the world-recognized most economical car on the highways—the Ace 400ZT.
>
> Designer jeans, polo shirts, a dazzling array of blouses, dresses, skirts, and accessories are waiting for you.

Fourth, present evidence of your product or service's application. Emphasize its convenience, usefulness, and economy. Endorsements, guarantees, and special features may be highlighted. It is here that you present the facts, but do so in a manner that emphasizes the attractiveness of your offer.

Examples

> A member of Time-Share International, Driftwood Resort offers not only the most reasonable prices on the Gold Coast but also opportunity to vacation in 39 countries of the world. Spacious two and three bedroom plus studio accommodations are available to suit your precise requirements. Fully equipped kitchens with modern hottubs allow you the casual lifestyle of a truly refreshing holiday.
>
> Fifty-three motoring journalists from 15 European nations voted this newest Ace "Car of the Year." The 400ZT is aerodynamically designed to accelerate from zero to fifty in only seven seconds. Disc brakes, rack and pinion steering, and a performance-tuned suspension system make this automobile a marvel to drive.
>
> College Corner clothes will make you feel confident and poised. Our College Board representatives make sure we buy the "in" ensembles to fit your classroom, party, and extracurricular needs. We even offer free alterations.

Finally, urge your reader to action. Make it easy for your readers to return a postcard to order your product, suggest an appointment next week, invite them into your store for a free gift, include a phone number to call, and so on.

Examples

> Call 305-455-9000 to arrange a tour of our facilities and to find out what valuable gift is yours. You won't be disappointed.
>
> Stop by this weekend to test drive your next car—the Ace 400ZT.

> During Orientation Week we will be open until 9:00 P.M. for you to drop in and browse. Free textbook covers are yours with every purchase.

Figure 2.12 shows a persuasive service offer letter. The audience consists of busy top executives who may be frustrated by the poor quality and time-consuming writing skills of their employees. The layout is catchy and appealing. The service's application and the desired action is effectively covered, yet the letter is brief.

Employment Application and Cover Letters

Because employment application letters often include a resume, a separate writing strategy, they will be discussed separately in Chapter 3.

Transmittal Letters

Letters of transmittal announce the enclosure of attached material and reports. Content and samples are covered in Chapters 3 and 6.

BAD NEWS LETTERS

Some letters must, of necessity, convey "bad news." They may inform the recipients that they are not hired, cannot get a refund, are late with a payment, and so on. Again there are writing strategies to convey your "bad news" in a positive, result-producing manner. "Bad news" letters include

- Negative response letters
- Complaints
- Collection letters
- Solicitations

Negative Response Letters

A letter which must say *no* requires a buffer statement before the "bad news." A buffer statement presents a valid reason before the negative response.

Examples

> We lease our apartments only through registered real estate brokers; therefore, . . .

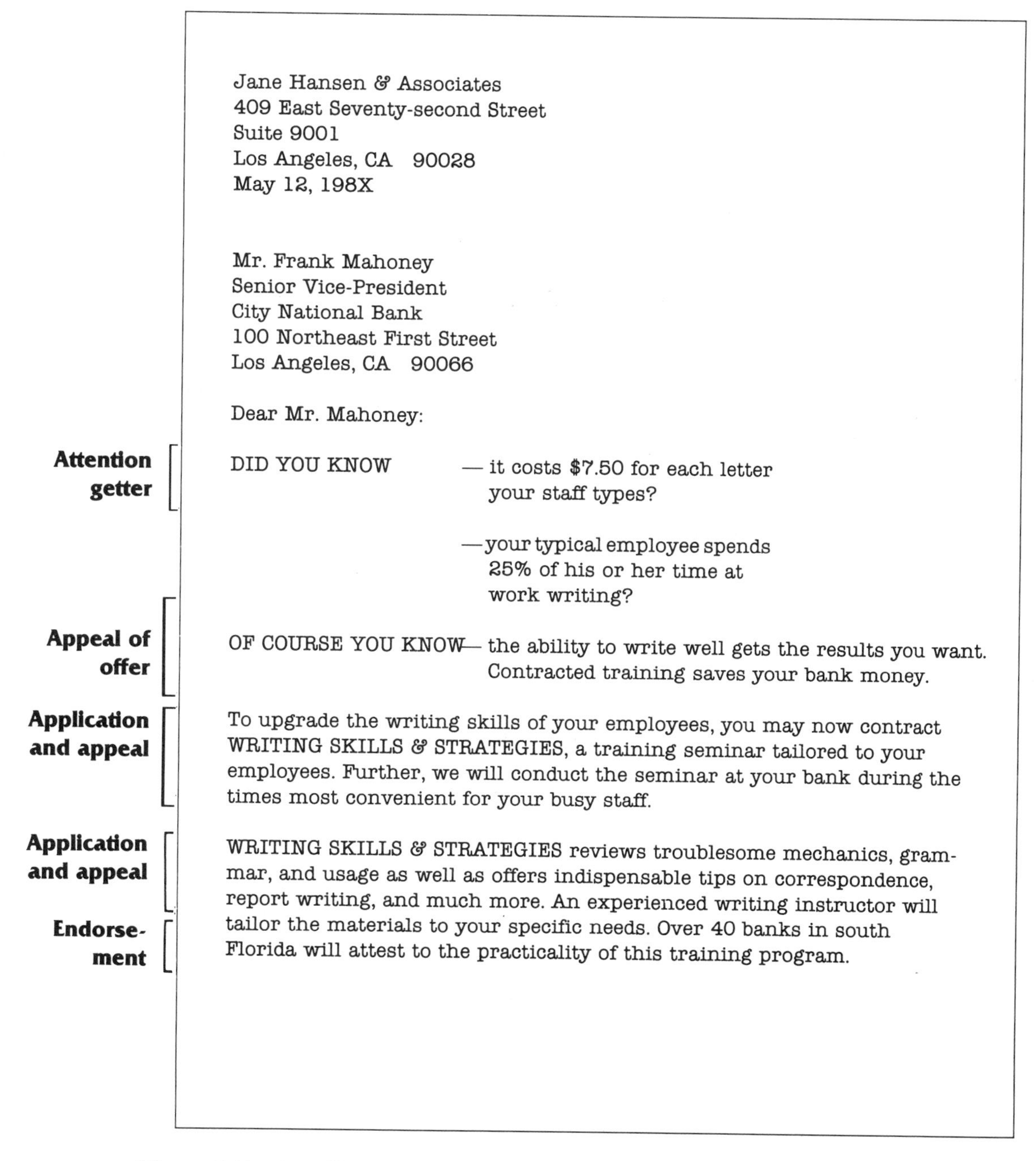

Jane Hansen & Associates
409 East Seventy-second Street
Suite 9001
Los Angeles, CA 90028
May 12, 198X

Mr. Frank Mahoney
Senior Vice-President
City National Bank
100 Northeast First Street
Los Angeles, CA 90066

Dear Mr. Mahoney:

DID YOU KNOW — it costs $7.50 for each letter your staff types?

— your typical employee spends 25% of his or her time at work writing?

OF COURSE YOU KNOW— the ability to write well gets the results you want. Contracted training saves your bank money.

To upgrade the writing skills of your employees, you may now contract WRITING SKILLS & STRATEGIES, a training seminar tailored to your employees. Further, we will conduct the seminar at your bank during the times most convenient for your busy staff.

WRITING SKILLS & STRATEGIES reviews troublesome mechanics, grammar, and usage as well as offers indispensable tips on correspondence, report writing, and much more. An experienced writing instructor will tailor the materials to your specific needs. Over 40 banks in south Florida will attest to the practicality of this training program.

Figure 2.12 An effective service offer letter

Urge to action

May I make an appointment, Mr. Mahoney, to discuss course content, prices, and times? I will call your office within the next ten days.

Very truly yours,

Jane Hansen

Jane Hansen
Writing Consultant

JH:jsv

Application

P.S. The enclosed brochure highlights features of WRITING SKILLS & STRATEGIES.

Figure 2.12 Continued

> Because the criteria for the position of Personnel Specialist requires a college degree in business administration, we can consider only those applicants with that credential.
>
> I have referred your letter to Ms Jane Clifford, assistant to the director of computer services, because only that department is authorized to give you the information you request.

A second strategy is to avoid the word *no.*

Too harsh I'm sorry to say no, I cannot address your engineers next week.

Buffered Because I will be in Chicago all of July, I cannot speak before your group.

A third strategy in a negative response letter is to avoid an apology. Your reasons are valid, so eliminate *unfortunately, we regret, we are sorry, we wish we could,* and so on.

Fourth, do not leave the opportunity to reopen discussion or consideration. Avoid "If you wish to discuss this further . . . " and "We wish we could . . . "

Finally, conclude by establishing goodwill. Use "We hope we have an opportunity to serve you in the future," "We appreciate your interest in our organization," and the like. Sincerity is an important consideration. If your effort to establish goodwill appears contrived, forced, or formulistic, it will irritate your reader.

Figure 2.13 illustrates an effective negative response letter to a job applicant. It buffers the bad news, does not apologize, and establishes goodwill.

Complaints

All of us have found it necessary to complain about defective products, delayed orders, billing errors, or inadequate services. Although we usually write complaint letters at a time of anger or frustration, angry tones seldom elicit the action we desire. A complaint letter needs restraint, specificity, and a clear statement of desired action.

The first strategy is to provide a detailed description of the faulty product, service, or suspected error along with the specifics about your purchase or contract.

Example

> I am returning for a full refund the U.D.S. Computer Telephone, model 333, for which I sent money order 40920 in the amount of $10.00 on July 15, 198X. I received the defective phone on August 20, 198X.

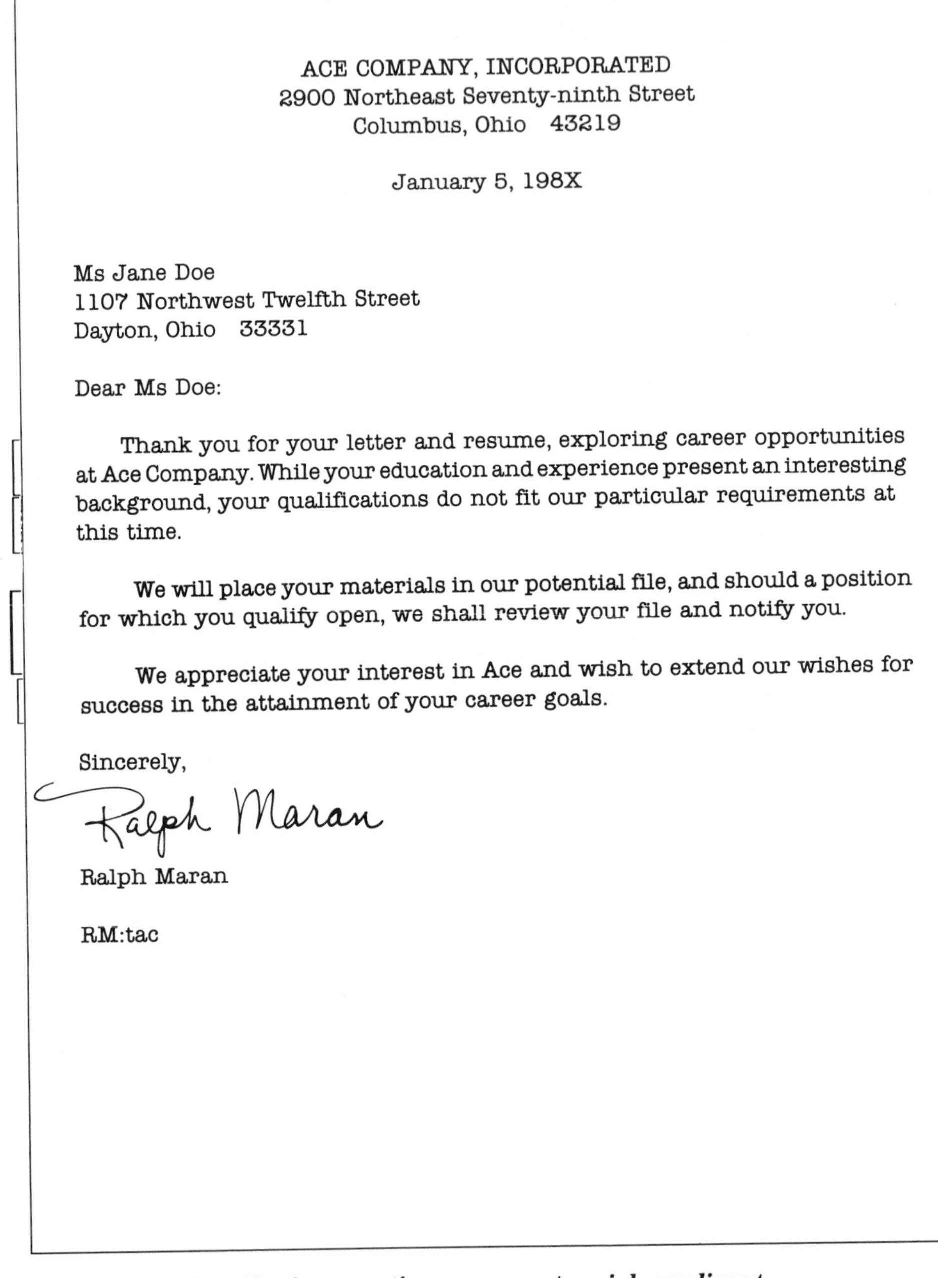

ACE COMPANY, INCORPORATED
2900 Northeast Seventy-ninth Street
Columbus, Ohio 43219

January 5, 198X

Ms Jane Doe
1107 Northwest Twelfth Street
Dayton, Ohio 33331

Dear Ms Doe:

Thank you for your letter and resume, exploring career opportunities at Ace Company. While your education and experience present an interesting background, your qualifications do not fit our particular requirements at this time.

We will place your materials in our potential file, and should a position for which you qualify open, we shall review your file and notify you.

We appreciate your interest in Ace and wish to extend our wishes for success in the attainment of your career goals.

Sincerely,

Ralph Maran

Ralph Maran

RM:tac

Figure 2.13 An effective negative response to a job applicant

Secondly, state precisely what is wrong with the product or service.

Example

The phone malfunctions. The beeper activates on the third dialed digit so that dialing cannot be completed. A persistent, loud buzzing interferes with reception on incoming calls.

Third, consider describing the inconveniences you have experienced. This is not always necessary but may help to underscore the seriousness of your complaint.

Example

Because I conduct a great deal of business from my home telephone, I have lost sales and commissions by not having a properly functioning instrument.

Finally, state clearly what action you desire. You may want a refund, a replacement, copies of all records, or some other consideration.

Example

I am enclosing the warranty and am requesting a full refund for the purchase price. I will not consider a replacement because I have lost confidence in your merchandise. I shall appreciate a prompt refund check.

Figure 2.14 illustrates a complaint about a rebate offer.

Collection Letter

Unfortunately, not all consumers pay their bills on time. Therefore, companies must employ several correspondence strategies to urge payment. Frequently, companies send a series of collection letters, each employing a stronger tone than the former. It is not wise to demand immediate payment in the first collection letter because there may be valid reasons for slow payment, such as misdelivered or misplaced bills, errors in the company's billing, and so on. Further, an early threat to begin legal action may cause the well-intentioned customer to avoid further profitable transactions with the company, or the company may suffer negative, word-of-mouth publicity. A tactful "you" perspective is important in maintaining good relationships.

The first letter should make the customer feel valuable, allow the customer to save face, urge prompt payment, but offer to establish a partial payment schedule.

8625 West Ninety-sixth Street
Trenton, New Jersey 08618
July 9, 198X

Service Merchandise Company, Inc.
1605 South Federal Highway
Trenton, New Jersey 08628

Gentlemen:

Specifics of purchase

On May 20, 198X, I purchased a General Electric smoke alarm, model 702, from your store. A $5.00 rebate, which I have not received, was offered with the product.

Specifics of complaint

Because the store did not have the proper form for requesting the rebate from General Electric, your clerk instructed me to send the original receipt and alarm box showing the model number to the manufacturer. I mailed both on May 22. I have received a letter from General Electric, dated July 5, 198X, signed by David O'Malley, stating that the rebate is no longer being offered.

Desired action

Would you please call David O'Malley in the Customer Relations department of General Electric, 205-907-4242, and explain (1) I purchased the alarm while the rebate offer was in effect, (2) your store was remiss in not having the rebate claim form on hand, and (3) I should be paid the $5.00 rebate promptly.

I shall appreciate your handling my problem as quickly as possible.

Yours truly,

Nicholas Anthony

Nicholas Anthony

NA:mt

Figure 2.14 An effective complaint letter

"You" perspective buffer

"You" perspective urge for full payment

Urge for action

ACE DEPARTMENT STORE
491 East Seventh Way
Peoria, Illinois 61650

August 1, 198X

Mrs. Quentin Adler
2201 West Orange Boulevard
Oak Lawn, Illinois 60454

RE: ACCOUNT #427-010-20

Dear Mrs. Adler:

Ace Department Store values you as a shopper. Offering you a charge account is just one way we try to make your shopping more convenient.

Our records indicate that you have missed two payments in June and July and that your amount past due as of August 1, 198X, is now $420.00. Unless you have a question regarding this figure, we would appreciate receiving your full payment. If you are unable to pay the full due amount now, won't you please call our Customer Service Department to set up a partial payment schedule in order to maintain your credit with us?

We will appreciate your prompt attention to this matter.

Very truly yours,

Marian Williams

Marian Williams
Customer Service

MW:jn

Figure 2.15 An effective initial collection letter

A follow-up letter should refer to the first request for payment and urge partial or full payment by return mail in order to avoid "further action."

A final letter usually refers to the first requests for payment and alerts the customer that the account will be turned over to a collection agency if partial or full payment isn't made immediately.

Figure 2.15 illustrates an initial letter of collection to a customer for an overdue charge account payment.

Solicitations

Although not truly "bad news," a solicitation for money or volunteer activities may be classed here because such solicitations offer nothing in return except, perhaps, the opportunity to further a cause which interests the reader. Alumni groups, political candidates, supporters of various causes, and the like frequently solicit money and volunteers. Such letters require inventive, eye-catching, persuasive techniques. Among these techniques are those used in advertising: bandwagon appeal, snob appeal, humor, endorsements, and so on.

The humorous approach is one effective technique, as illustrated in Figure 2.16, a letter soliciting a contribution from a college alumnus.

the annual program of THE MIAMI UNIVERSITY FUND
OXFORD, OHIO 45056 • TELEPHONE (513) 529-5211

Projects for 1983:

Scholarships and Student Aid
The University Library
International Studies Program
Academic Advancement
Undergraduate and Graduate Research
The Art Museum
Sesquicentennial Chapel
Miami University Student Foundation
Cultural Enrichment and Other Activities
Environmental Sciences
Formal Gardens
Alumni and Development Activities
Men's and Women's Athletics — "The TRIBE"

June, 1983

YOU NEVER KNOW
WHERE AN "OSCAR" LURKS:

Take Clark Gable and his first Oscar.

It sprung from a hidden opportunity, which was followed by dramatic consequences.

It seems that Clark was getting a little uppity at mighty MGM, back there is 1934, so they "loaned" him to run-down, poor-relation Columbia Pictures for a film.

That was like sending a Concorde captain out to dust crops in Kansas.

Clark was mad. Didn't want to go. Resisted. Had no choice.

Eight leading actresses turned down the chance to star with Gable. They just couldn't stomach the script. Miriam Hopkins offered to buy it and throw it away rather than play in it.

Really.

Claudette Colbert needed the money; had four weeks vacation; offered to hold her nose and share Clark's punishment if they promised to finish it fast. Frank Capra sweet-talked the two of them through It Happened One Night.

many gifts combine to do great things for Miami

Figure 2.16 An effective solicitation letter (Courtesy of Douglas M. Wilson)

It turned out to be the first film ever to sweep the major Oscars,

. . . best screenplay,

. . . best picture,

. . . best director,

. . . best actor,

. . . best actress.

Wow!

Just proves you never know when Opportunity will knock.

Same way with The Miami University Fund.

You knew I'd get to it, didn't you?

Well, the money you send the Fund certainly provides opportunities for many . . . through scholarships, library books, cultural programs, plus many more.

And often the consequences are dramatic.

The difference between these opportunities and Gable's, of course, is that everyone's happy to get them, even though no one ever really knows what will happen as a result.

But because the tens of thousands of gifts, small and large, add up to millions of dollars each year, we do know that the fund makes the whole Miami experience better for all . . . attracts fine students . . . improves the faculty . . . even reflects favorably on Alumni, who make it possible.

This solid alumni support is a good part of what makes you, and all of us, proud to be connected with Miami, I'm sure.

So please don't neglect your "opportunity" to keep these standards high in 1983. It's the best way I know of to help keep Miami rated:

Figure 2.16 Continued

. . . best students

. . . best faculty

. . . best experience

. . . best university

. . . best Alumni

Doug Wilson

Douglas M. Wilson
Vice President
University Relations

DMW:fhb

Figure 2.16 Continued

EXERCISES

1. Rewrite the following memo subject lines to make them more precise and descriptive:

Test Results	Vacations
Minutes	Training
Meeting	New Personnel
New Policy	Copy Machine
Schedules	Hours

2. Correct the errors in the letter elements of the following block letter format.

Ms. Julie Wilson
2 N. W. Park Ave.
Chicago, Ill.   33302

Mr. Daniel C. Taylor
four-one-two E. 72nd St.
Dayton, Oh,   40727

Dear Sir,

Re: Account #407-201 E

Best,

Marcia Morris

Marcia Morris, Treasurer

mm:tc

3. Rewrite the body of the following complaint letter by dividing it into an introduction, body, and closing and by eliminating "letterese," a "me" perspective, and an inappropriate tone.

> Dear Sir:
>
> In reference to your lousy iron which I purchased recently, I want my hard-earned money back. If you don't refund me the price in full, I beg to inform you that I will take legal action. It spews water all over the clothes I iron, scorches things even on a low setting, will not stand up on its base, and the plug broke the last time I plugged it in. If you have any questions, do not hesitate to call me. I am appalled at the workmanship of this piece of junk. Get with it.
> Cordially,

WRITING OPTIONS

1. Write a memo to the people with whom you work (real or imaginary) announcing a meeting for a specific purpose. Be very specific about date, place, time, and purpose.
2. Write a memo to your employer (real or imaginary) to point out a minor problem at your place of work, such as a scheduling mixup, inadequate lighting, the need for more storage shelves or files, your inability to perform an assigned task, an error in your paycheck, and so on. Be brief, to-the-point, and focused.
3. Write a "good news" letter.
   a. Write an inquiry or request letter to a company or other organization in response to an advertisement or article in a professional journal in your field. Ask at least four technical questions about a product or service. Review the writing strategies which will motivate a quick response.
   b. Write an order letter for merchandise from a particular company. Specify the stock number, quantity, size, color, unit cost, and total cost as appropriate. Include all details about payment and delivery.
   c. Write a congratulatory or thank-you letter. Some suggested topics are congratulations to a friend who has been hired or promoted, won a professional award, completed a degree or other special training, been elected to public or organizational office, or opened up a business. Thank-you letter topics might be a response to a letter of recommendation, a letter of appreciation to a teacher or counselor whose advice you followed, to a hotel

or restaurant which hosted a group function or dinner, or to a person who directed some business opportunity your way.

d. Write a sales or service offer letter. Consider your audience's age, occupation, geographical location, needs, and interest. Some suggested topics are the merits of a particular automobile, college, restaurant, new store, personal computer, bank, travel bureau, flower shop, or maintenance service for lawn, pool, or snow removal.

4. Write a "bad news" letter.

a. Write a negative response letter. Some suggested topics are turning down an offered job, declining an invitation to speak, declining to serve on a committee or joining an organization, refusal to volunteer time to an organization, refusal to a job applicant, inability to fulfill a reservation request at a hotel or travel group, declining to refund money for a specific piece of merchandise, or refusal to a sales or service offer.

b. Write a complaint letter. Some suggested topics are an error in your credit card or telephone bill, rude service you received at a store or restaurant, late delivery of merchandise, poor quality of some product recently purchased, or damaged goods delivered by bus, air, or rail.

c. Write a series of three collection letters, each firmer than the last. Some suggested topics are late dues payment to a club or organization, late payment for a service you rendered, or late payment for an automobile, credit account, or bank loan.

d. Write a solicitation letter. Select a cause which interests you—save the whales or manatees, abortion legislation, a political candidate, the equal rights amendment, zoning irregularities, a halfway house for troubled youth, and so on—and solicit donations or appearance at a civic forum to discuss the issue with top policymakers.

3

RESUMES, COVER LETTERS, AND INTERVIEWS

Skills:

After studying this chapter, you should be able to

1. Define *resume* and *cover letter.*
2. Appreciate the importance of a detailed, letter-perfect resume and cover letter.
3. Compile a personal dossier of educational and employment background materials and reference letters.
4. Compile a dossier of future position prospects.
5. Write a personal resume in an appropriate format, including career objectives, personal data, educational background, employment history, and references.
6. Write a cover letter to a specific prospective employer.

INTRODUCTION

Résumé is a French word which means "summary." A personal resume is a concise summary of pertinent facts about yourself—your employment objectives, your employment and educational history, personal data, and reference lists. Although most dictionaries include the accent marks (résumé), it is common practice to omit them. A resume and its accompanying cover letter are indispensable job-hunting tools which are submitted to employers to "sell" yourself as a prospective employee and to obtain an interview.

Your resume allows you to organize and amplify your data in a manner which highlights your strengths. A job application form (see Figure 3.1) usually restricts you to a mere listing of schools and previous employment names and addresses. Although the resume and cover letter are usually mailed to companies who have advertised for applicants or to those companies where you know an opening exists, you may present a resume during an interview which you have obtained by other means. The summary provides a focal point for discussion between you (the applicant) and the interviewer.

RECORD KEEPING

Long before you seek initial employment in your chosen career or seek to change jobs, it is wise to compile a file of your achievements and

APPLICATION FOR EMPLOYMENT

The Civil Rights Act of 1964 prohibits discrimination in employment because of race, color, religion or national origin.
Public Law 90-202 prohibits discrimination because of age.

PERSONAL INFORMATION

DATE — SOCIAL SECURITY NUMBER

NAME: LAST — FIRST — MIDDLE — AGE — SEX

PRESENT ADDRESS: STREET — CITY — STATE

PERMANENT ADDRESS: STREET — CITY — STATE

PHONE NO. — OWN HOME — RENT — BOARD

DATE OF BIRTH — HEIGHT — WEIGHT — COLOR OF HAIR — COLOR OF EYES

MARRIED — SINGLE — WIDOWED — DIVORCED — SEPARATED

NUMBER OF CHILDREN — DEPENDENTS OTHER THAN WIFE OR CHILDREN — CITIZEN OF U.S.A. YES ☐ NO ☐

IF RELATED TO ANYONE IN OUR EMPLOY, STATE NAME AND DEPARTMENT — REFERRED BY

LAST — FIRST — MIDDLE

EMPLOYMENT DESIRED

POSITION — DATE YOU CAN START — SALARY DESIRED

ARE YOU EMPLOYED NOW? — IF SO MAY WE INQUIRE OF YOUR PRESENT EMPLOYER?

EVER APPLIED TO THIS COMPANY BEFORE? — WHERE — WHEN

EDUCATION	NAME AND LOCATION OF SCHOOL	YEARS ATTENDED	DATE GRADUATED	SUBJECTS STUDIED
GRAMMAR SCHOOL				
HIGH SCHOOL				
COLLEGE				
TRADE, BUSINESS OR CORRESPONDENCE SCHOOL				

SUBJECTS OF SPECIAL STUDY OR RESEARCH WORK

WHAT FOREIGN LANGUAGES DO YOU SPEAK FLUENTLY? — READ — WRITE

U.S. MILITARY OR NAVAL SERVICE — RANK — PRESENT MEMBERSHIP IN NATIONAL GUARD OR RESERVES

ACTIVITIES OTHER THAN RELIGIOUS (CIVIL, ATHLETIC, FRATERNAL, ETC.)

(CONTINUED ON OTHER SIDE)

Figure 3.1 Sample employment application

FORMER EMPLOYEES (LIST BELOW LAST FOUR EMPLOYERS. STARTING WITH LAST ONE FIRST)

DATE MONTH AND YEAR	NAME AND ADDRESS OF EMPLOYER	SALARY	POSITION	REASON FOR LEAVING
FROM TO				
FROM TO				
FROM TO				
FROM TO				

REFERENCES: GIVE BELOW THE NAMES OF THREE PERSONS NOT RELATED TO YOU, WHOM YOU HAVE KNOWN AT LEAST ONE YEAR.

NAME	ADDRESS	BUSINESS	YEARS ACQUAINTED
1			
2			
3			

PHYSICAL RECORD:
LIST ANY PHYSICAL DEFECTS

WERE YOU EVER INJURED? GIVE DETAILS

HAVE YOU ANY DEFECTS IN HEARING? IN VISION? IN SPEECH?

IN CASE OF EMERGENCY NOTIFY

NAME ADDRESS PHONE NO.

I AUTHORIZE INVESTIGATION OF ALL STATEMENTS CONTAINED IN THIS APPLICATION. I UNDERSTAND THAT MISREPRESENTATION OR OMISSION OF FACTS CALLED FOR IS CAUSE FOR DISMISSAL. FURTHER, I UNDERSTAND AND AGREE THAT MY EMPLOYMENT IS FOR NO DEFINITE PERIOD AND MAY, REGARDLESS OF THE DATE OF PAYMENT OF MY WAGES AND SALARY, BE TERMINATED AT ANY TIME WITHOUT ANY PREVIOUS NOTICE.

DATE SIGNATURE

DO NOT WRITE BELOW THIS LINE

INTERVIEWED BY DATE

REMARKS:

NEATNESS		CHARACTER	
PERSONALITY		ABILITY	

HIRED FOR DEPT: POSITION WILL REPORT SALARY WAGES

APPROVED 1. 2. 3.

EMPLOYMENT MANAGER DEPT. HEAD GENERAL MANAGER

This Application form is made for general use and distribution in United States and the manufacturer cannot assume responsibility for the inclusion in said form of any questions which may be at variance with local and state laws.

Figure 3.1 Continued

employment experiences. This file, or dossier, should also contain all leads you can gather on employment possibilities and a list of personal references. A complete dossier will make your writing tasks easier.

For your file, compile a list of the names and addresses of all educational institutions you have attended (high schools, community colleges, universities, trade and vocational schools, and military schools). Note the inclusive dates (months and years) of your attendance. List your degrees, fields of study, outstanding awards, achievements, and grades. Also include records of your employment history: company names, addresses, position titles, dates, and summaries of your responsibilities.

List names, addresses, and phone numbers of prospective references. These may include present and past employers, teachers and professors of courses in your chosen field, recognized leaders in your field with whom you have been in work-oriented contact, clergy, and other professional and experienced men and women who can attest to your skills, abilities, and personal attributes.

Each time you obtain a new skill, such as mastering a word-processing method, earning a certificate for achieving new typing or dictation skills, learning an emergency medical rescue technique, earning a new license, attending a management workshop or seminar, and so on, drop a note into your file.

Also include in your file a list of all prospective employers. You should be active in searching out all of the companies, agencies, or institutions which might have future openings for which you qualify. Some sources for the job search include

1. **Help-wanted Columns.** Read your newspaper classified section to determine the openings in your field. Note the number of openings, the job descriptions, the qualifications, and salary ranges. List the companies and details.
2. **Classified Telephone Directories.** Thumb through the Yellow Pages and list the names of companies or agencies which might be prospective employers.
3. **Civil Service Offices.** Call federal, state, county, and city civil service offices. These offices list government positions in such fields as engineering, building and construction, technology, law enforcement, and parks and recreation. Some offices offer prerecorded recruitment telephone tapes, listing open positions. Others print lists of local, state, and national positions along with qualifications, salaries, and application details.
4. **Trade Magazines and Newspapers.** Buy or subscribe to magazines or newspapers which specialize in your field. These publications often include Help Wanted sections.
5. **Libraries.** Ask the reference librarian to help you locate occupational handbooks, government publications, and newsletters which contain information on openings and qualification requirements in your field.

6. **College Placement Offices.** Visit your college or university career service center. Most offices offer interest inventory tests, career counseling, computer-assisted career guidance programs, occupational briefs and recruitment brochures, and job files of local, state, and national full- and part-time positions.
7. **Interviews.** Actively arrange to speak to faculty in your major field and people already employed by companies which interest you. They can offer invaluable, practical advice and employment tips.
8. **Employment Agencies.** Many companies do not advertise openings, but put their employment search tasks in the hands of employment agencies and executive search companies. Some agencies specialize in certain types of employment, such as electronics, money and banking, allied health fields, and so on. Incidentally, employers often pay the fee to the agency at no cost to you. Seek those agencies which can help you.

In summary, be informed and realistic about your job prospects. Learn how your qualifications fit the needs of prospective employers.

THE RESUME

Format

Although there are several acceptable formats for an effective resume, all of them are divided by similar sections to provide a complete inventory of your qualifications and experience. For easy preparation and reading, try to contain your information on no more than two pages. The sections will include

- name, address, and phone
- career objectives
- educational background
- employment experience
- personal data
- references

Employers skim resumes, sorting out those which are uninteresting or messy. Therefore, you want to present a letter-perfect resume which, due to overall layout, headings, and white space, not only is readable but also quickly highlights your strengths. Copies should be quick-printed on bond paper rather than photocopied. Figure 3.2 shows a sample resume format. Headings may be centered or placed flush left to the margin. The bracketed items are optional, depending on the position for which you are applying.

Resume of JANE C. DOE	Street address City, State Zip Phone: 000-000-0000

OBJECTIVE

Name of position or broader statement indicating long-term objective

EDUCATION

Month, year to Present	Name of present institution Street address City, State Zip • Phrase on degree sought, major, and expected date of graduation • [Several bulleted listings of high accumulated grade point, honors, activities, organizations, certificates earned]
Month, year– Month, year	Previous schools Street address City, State Zip • Types of degrees, diplomas, major achievements
Dates	[Miscellaneous educational experiences, such as company courses, correspondence courses, seminars, home study]

EMPLOYMENT

Month, year to Present	Name of company Street address City, State Zip • Job title • [Several bulleted phrases which amplify the duties performed, promotions, awards, and, possibly, the reason for leaving]

Figure 3.2 Sample resume format

Resume of
JANE C. DOE

page 2

Month, year– Month, Year	Name of company Street address City, State Zip • Job title • [Several bulleted phrases of amplification]

PERSONAL

Age:	years [Day, month, year; place of birth]
[Appearance:	Height: Weight:]
Marital Status:	Status [If married, spouse's name, number and ages of children]
[Health:	State of, significant limitations; date of last check-up]
[Citizenship:	Country; work visas]
[Military Status:	Rank; Service; date of discharge]
[Residence:	Length of state residence; own, rent, live with parents; will or will not relocate]
[Affiliations:	Other than religious, list community, fraternal, etc.]
[Special interests:	Travel abroad; languages, skills, *no sports*]

REFERENCES

References available upon request [if sending resume to large number of employers]

or

Name, Title Company or Institution Street address City, State Zip Phone: 000-000-0000	Name, Title Company or Institution Street address City, State Zip Phone: 000-000-0000

Figure 3.2 Continued

Resume of
JANE C. DOE

page 3

Name, Title
Company or Institution
Street address
City, State Zip
Phone: 000-000-0000

Name, Title
Company or Institution
Street address
City, State Zip
Phone: 000-000-0000

Figure 3.2 Continued

Names and Addresses

In the upper left-hand corner type "Resume of" and your full, legal name. In the upper right-hand corner type your street address, city, state, zip code, and phone number. Include your area code. You may wish to include both your home and business phones. See Figures 3.2, 3.3, and 3.4 for examples.

Job Objective

Center the heading "Objective" in capital letters and underline it. Beneath it type the exact desired position you are seeking or a brief phrase which spells out your short- and long-term career objectives. Avoid full sentences.

Examples

Legal Secretary

Position in Customer Service or Sales Field

Position in drafting leading to design responsibilities

Electronic Technician with opportunity for advancement into management

Entry-level position in data processing leading to computer programming

Educational Background

If your educational preparation is stronger than your employment experience, develop this section first. Type the heading "Education." Beneath it begin with your most recent school and list all other schools in reverse chronological order. List the inclusive dates, including the months, in one column. In another column include the names of the institutions, addresses, cities, states, and zip codes. Cite your major field of study, degrees earned or expected, and graduation dates. Highlight your academic record (if high), awards, special activities, organizations, special skills, and job-related courses. Include only those achievements, however, which are slanted towards the job. A prospective employer in banking might be impressed if you were treasurer of your student government association, but would not care if you were on the tennis team.

Employment Experience

If your previous employment is more indicative of your qualifications, place this section before your educational data. Type the heading "Employment" or "Work Experience" in capital letters. Beneath it begin with your present or most recent employment and work backward in time. Include the inclusive dates, including months, in one column. Employers are often looking for gaps in both education and employment histories. A continuous record of employment or education suggests that you are a responsible individual with work-oriented goals. In another column include the names of the companies, addresses, cities, states, zip codes, and phone numbers. List your position titles and whether the work was part- or full-time. Using brief phrases ("Responsible for . . . ," "Implemented . . . ," "Maintained . . . "), highlight your duties, achievements, and awards. You may want to list your reason for leaving ("Offered higher-paying position," "Returned to college full-time").

Personal Data

Type the heading "Personal." Consider what data your prospective employer wants to know. At the least include your age, marital status, and general health status. Decide whether to include your place of birth, length of local residency, ownership of home, relocation willingness, military service, community affiliations, and special skills and interests other than sports. You may list these in tabular form as shown in Figures 3.3, 3.4, and 3.5.

References

Type the heading "References." If you are submitting your resume to a large number of employers, it will be wise to write only "References will be furnished upon request." Otherwise each reference may be unduly bothered by spot checks from many companies. If you are selective about your application, list four or five people who can supply strong, positive assessments of your qualifications and character. Obtain permission in advance to list anyone as a reference. Select present or past employers, college instructors, clergy, and other community leaders. Do not include relatives. Type the name, position, company affiliation, street address, city, state, zip code, and phone number of each reference.

Figures 3.3, 3.4, and 3.5 illustrate sample resumes. Figure 3.3 is a resume of a typical college student earning an associate degree in an allied health field. Figure 3.4 is a resume of a mature student with wider experience and training than a recent college graduate. Figure 3.5 illustrates an alternative format.

Cover Letter

A cover letter, also called a *letter of application* or a *face letter,* accompanies your resume and serves both as an introduction of yourself and as a strategy to interest employers sufficiently to read your resume and to

Résumé of
JOANNE L. FATA

11701 Southwest Tenth Place
Fort Lauderdale, Florida 33325
305-475-9641

OBJECTIVE

Medical Laboratory Technician at a modern, expanding hospital laboratory.

EDUCATION

August 1977 – Present

Broward Community College
Central Campus
3501 Southwest Davie Road
Fort Lauderdale, Florida 33314

- Majoring in Medical Laboratory Technology; will receive Associate in Science Degree
- Grade-point average: 3.1 of 4.0
- Special Courses: Advanced Instrumentation, Advanced Medical Laboratory Techniques, Advanced level of Mycology

November 1976 – November 1977

Sheridan Vocational Center
5400 Sheridan Street
Hollywood, Florida 33021

- Completed the NAACLS accredited 12-month Certified Laboratory Assistant program

EMPLOYMENT

August 1979 – Present

Florida Medical Center
5000 West Oakland Park Boulevard
Fort Lauderdale, Florida 33313
305-735-6000

- Laboratory Technician in the Hematology Department

Figure 3.3 Sample resume of student (Courtesy of student Joanne L. Fata)

Résumé of JOANNE L. FATA

Page 2

- Responsible for early morning start-up procedures on the Coulter Model S-Plus, Model S, Coulter Diff 3 System, and Coag-A-Mate
- Perform routine Urinalyses and Serology procedures
- Experienced in Whole Blood Calibration procedures
- Experienced in patient blood-drawing procedures

February 1978 – April 1979

Robert G. Talley, M.D. P.A.
912 East Broward Boulevard
Fort Lauderdale, Florida 33301
305-463-5271

- Responsible for all laboratory procedures including manual chemistry tests for Glucose, BUN, SGOT, Alkaline Phosphatase, Uric Acid, and Cholesterol
- Performed routine urinalysis, manual CBC's, and differentials
- Reason for leaving: Registered for full-time courses for three months

PERSONAL

Age: 24 Health: Excellent Marital Status: Single
Affiliation: American Society of Clinical Pathologists
Willing to relocate

Figure 3.3 Continued

Résumé of
JOANNE L. FATA

Page 3

REFERENCES

Ms. Rebecca Meehan, MT (ASCP)
Chief Laboratory Technologist
Florida Medical Center
5000 West Oakland Park Boulevard
Fort Lauderdale, Florida 33313
305-735-6000

Robert G. Talley, M.D. P.A.
912 East Broward Boulevard
Fort Lauderdale, Florida 33301
305-463-5271

Ms Karen Mead, MT (ASCP)
Hematology Supervisor
Florida Medical Center
5000 West Oakland Park Boulevard
Fort Lauderdale, Florida 33313
305-735-6000

Ms Barbara L. Kremp, MT (ASCP)
Broward Community College
Allied Health Department
Central Campus
3501 Southwest Davie Road
Fort Lauderdale, Florida 33314
305-581-8700

Figure 3.3 Continued

<table>
<tr><td>Résumé of
SANDRA J. BURNETTE</td><td>7661 Hood Street
Hollywood, FL 33024
305-962-2365</td></tr>
</table>

OBJECTIVE

Seeking an entry-level position in your data processing department leading to a computer programming position

PERSONAL

Date of birth: June 5, 195X	Health: excellent
Marital status: married, no children	Florida residency: 5 years

EDUCATION

<table>
<tr><td>January 198X
to Present</td><td>Broward Community College
Central and South Campuses
225 East Las Olas Boulevard
Fort Lauderdale, Florida 33301
• Majoring in Data Processing
• Will earn A.S. degree in August 1984
• GPA: 4.0 of possible 4.0
• Major courses: Programming in Basic, Programming in Cobol, Occupational Writing, Intermediate Algebra, Fundamentals of Data Processing
• Clep credit: Introductory Accounting I and II
• Awards: President's List</td></tr>
<tr><td>August 198X
to January 198X</td><td>Sheridan Vocational Center
5400 Sheridan Street
Hollywood, FL 33020
• Completed Bookkeeping Systems Program
• Completed a two-semester course in one semester
• Major courses: Advanced Accounting, Microcomputer applications</td></tr>
</table>

Figure 3.4 Sample resume of mature student (Courtesy of student Sandra J. Burnette)

Résumé of
SANDRA J. BURNETTE

page 2

September 196X to June 197X

Cardinal High School
Thompson Avenue
Middlefield, OH 44062

- GPA: 3.89 of possible 4.0
- Major courses: Bookkeeping, Typing, Business English and Communications, Shorthand
- Awards: Valedictorian, National Honor Society, Scholastic Team

EXPERIENCE

February 198X to July 198X

D&J Construction, Incorporated
5155 Southwest Sixty-fourth Avenue
Davie, FL 33315
305-587-2360

- Maintained sales, cash receipts, cash payments journals, accounts payable sub-ledger
- Completed weekly payroll for sixty employees
- Completed associated quarterly payroll tax forms
- Maintained personnel and group insurance records for sixty employees
- Filed workers' compensation incident reports
- Prepared quarterly reports and summaries for use by accounting firm
- Conducted interviews for my replacement
- Trained my replacement

Figure 3.4 Continued

Résumé of
SANDRA J. BURNETTE

page 3

August 197X to July 198X	Southern Door Company 3600 North Twenty-ninth Avenue Hollywood, FL 33021 305-922-6747 • Assistant Bookkeeper • Maintained accounts receivable sub-ledger • Completed weekly payroll for forty employees • Paid miscellaneous accounts payable • Maintained raw materials inventory • Took customer phone orders • Performed customer service duties
197X to 197X	Secretarial and clerical positions for three insurance companies and a rubber manufacturing firm

REFERENCES

Mrs. Deanna J. Siegel
Controller
Southern Door Company
3600 North Twenty-ninth Avenue
Hollywood, FL 33021
305-922-6747

Mr. Frank Pittnaro
Certified Public Accountant
William Webb and Associates
400 East Atlantic Boulevard
Pompano Beach, FL 33060
305-921-6000

Mrs. Eileen Duffie
Bookkeeping Systems Teacher
Sheridan Vocational Center
5400 Sheridan Street
Hollywood, Fl 33020
305-963-8600

Mr. Jack Attaway
Vice-President
D & J Construction, Inc.
5115 Southwest Sixty-fourth Avenue
Davie, FL 33314
305-587-2360

Figure 3.4 Continued

Resume of
KENTON L. LAIRD

7061 Tyler Street
Hollywood, FL 33055
Hm: 305-962-2365
Bus: 305-472-1212

Objective

Electronic Technician in computer application company

Experience

June 198X to Present

Senior Field Engineer. Field Engineer of the Year Award. Maintained inventory control system for parts and complete units. Controlled expense system for service area. Assigned product responsibility for Beta and Gamma printers, VT-100, and Ace 5600 video terminals.
(Ace Communications Corporation, 201 South Johnson Street, Hollywood, Florida 33051)

June 197X to August 198X

Supervisor of Maintenance and Mechanical Equipment. Supervised 15 machinists. Charged with responsibility of keeping machinery and vehicles running in a plant employing 2000 on each shift. Products: electronic equipment used in modern medical laboratories.
(Greer Corporation, 4000 Palmetto Park Road, Miami, Florida, 31302)

May 197X to May 197X

United States Air Force. Maintained surveillance equipment on reconnaissance aircraft. At fighter base in Thailand assigned both aircraft and in-shop repairs; equipment included multi-channel tape recorders, radar receivers, signal analyzers, direction finders, and radio-jamming devices.

Figure 3.5 Sample alternate resume format

Resume of Kenton L. Laird continued Page 2

Education

June 198X to Present — Broward Community College
Fort Lauderdale, FL 33314

Seeking Associate in Science degree in Electronic Technology with a GPA of 3.7 out of possible 4.0. Expect to be graduated May 198X. Member of Association of Computing Machinery.

August 197X to May 197X — Keesler Air Force Base
Biloxi, MS 50001

Graduated with honors in Basic Electronics and Electronic Counter Measures courses.

August 197X to June 197X — Grant High School
Dayton, Ohio 70012

Personal

Age: 25
Languages: Fluent Spanish; Some Italian

Marital Status: Single
Military: Staff Sargeant, U.S. Air Force; Honorably discharged; May 198X

References

Business and personal references will be furnished upon request. DO NOT CONTACT PRESENT EMPLOYER AT THIS TIME.

Figure 3.5 Continued

grant you an interview. The letter should reflect your personality and emphasize your potential. While a model letter may be developed, each letter should be tailored to a particular prospective employer.

Each letter should be individually typed on bond paper of the same quality as your resume. Address your letter to a specific name, if possible.

Your opening paragraph must attract the reader's attention without being overly aggressive or "cute." You must indicate the position or type of employment you are seeking and refer to your enclosed resume. You may want to mention how you learned of the position and subtly praise the company. Consider these opening paragraphs:

> I am an ideal candidate for the electronics technician position which you advertised in the July 2 edition of The Miami Herald. My two years of experience in the field and an Associate of Science degree in Electronic Technology qualify me for employment with your progressive corporation. My enclosed resume amplifies my background.
>
> Mr. Jack Smith, Chair of the Data Processing Department at Broward Community College, has alerted me that there will be a computer technician opening in your computer maintenance group. As my resume details, my specific qualifications prove that I would be an asset to your organization.

Your body paragraph(s) should summarize and emphasize your specific qualifications and personal qualities as they relate to the position. Use an enthusiastic tone and project self-confidence. Mention your outstanding personal attributes. Consider this sample:

> I can offer you four years of experience in employment and education. I will earn my Associate of Science degree in radiology technology this May and have been employed as a medical assistant to Dr. James E. Perry, radiologist, here in Fort Lauderdale for two years. My personal attributes include determination, courtesy, and attention to detail.

The body material should be short, direct, and persuasive.

Your closing paragraphs should urge action on the part of the reader. Ask for an interview or suggest that you will phone shortly to arrange for an interview. If it is appropriate, seek additional information, applications, and so on. Mention your flexibility and availability. Consider these closings:

> May I have a personal interview to discuss your position and my qualifications? I will call your office on Monday to arrange a convenient date.
>
> I would welcome your consideration for an entry-level position and can start work any time next month. Any information you may have regarding my prospects with your company will be greatly appreciated. You may reach me after 3:00 P.M. weekdays at the number listed on my resume to arrange an interview.

Figures 3.6 and 3.7 (on pages 82–83) show the complete texts of sample cover letters.

INTERVIEWS

An interview is the final step in job hunting. Here your prospective employer evaluates you—your appearance, your personality, and your ability. Chapter 13 discusses a number of verbal strategies to employ in an interview, but a few tips are timely here.

1. **Be prepared.** In advance, learn as much as you can about the company so that you can project your interest and ask informed questions. Take a copy of your resume along to refresh your memory in response to direct questions about your background.
2. **Dress appropriately.** Although you will generally dress neatly and conservatively (a suit, a tie, a dark dress, hose and heels), wear some one article which is eye-catching and memorable. This might be a striped tie, a lapel rose, or an unusual, but not showy, piece of jewelry.
3. **Be on time.** Know the exact time and location of your interview. Arrive a few minutes early and state your name and purpose to the secretary, receptionist, or actual employer.
4. **Don't smoke.** You may be offered a cigarette, but refrain from accepting it. Spilled ashes, clouds of smoke, and ungainly stretches to an ashtray do not create a favorable impression.

7661 Hood Street
Hollywood, FL 33024
July 6, 198X

Mr. Jerry Diener
Personnel Director
American Express Company
777 American Expressway
Plantation, FL 33324

Dear Mr. Diener:

I am an applicant for an entry-level position in your data processing department. Dr. Ted Smith, head of the Data Processing Department at Broward Community College, South campus, has told me about your progressive practices, and I believe I can offer a substantial contribution to American Express Company. My resume highlights my qualifications.

My main experience is in the accounting field. To further my understanding of the complete accounting cycle, I enrolled in 198X at Sheridan Vocational Center in Hollywood, Florida. At that time I received my first exposure to computers. I found that I have a natural ability for the logical thinking that is required for a good computer programmer. Because a career in computer programming seems the ideal goal for me, I am now working towards my A.S. degree in data processing and will be graduated this August.

I will call you on Wednesday, July 13, to arrange an appointment for an interview.

Yours truly,

Sandra J. Burnette

Sandra J. Burnette

SJB

Enc: 1

Figure 3.6 Sample cover letter
(Courtesy of student Sandra J. Burnette)

5000 Griffin Road
Fort Lauderdale, FL 33314
May 2, 198X

Mr. J.W. Duran
Attorney-at-law
346 North Andrews Avenue
Fort Lauderdale, FL 33323

Dear Mr. Duran:

Please consider my application for your position of legal secretary which you advertised in the May 1 edition of the Fort Lauderdale News. I am the "trained professional who can perform a variety of office duties" you seek. My training in legal techniques, business law, and legal secretary practices plus three years of experience as a general secretary for Ace Construction Company definitely meet your qualifications. My complete resume is attached.

I will be graduated on June 9, 198X, from Broward Community College with an associate degree in Legal Secretarial Science. In addition to the required courses of this program, I have studied word processing software techniques. My shorthand rate is 120 words per minute, and I type at 65 words a minute.

In addition to specialized training, I can offer you three years of experience in general office work, bookkeeping, and salesmanship. You will find me to be reliable, efficient, and personable.

I trust you will consider me for your position. May I have a personal interview at your convenience? I may be reached by telephone between noon and 5:00 P.M. at 583-4771.

Very truly yours,

Sharon Cates

Sharon Cates

SC

Enclosure

Figure 3.7 Sample cover letter

EXERCISES

1. Using the following information, type the Employment section of a resume. Use a heading and arrange the data in a readable format. The applicant is seeking a position as an electronic technician with the opportunity for advancement. (1) His first job was as an auto mechanic with Bird Ford Agency, 101 North Federal Highway, Fort Lauderdale, Florida, where he serviced cars and was responsible for all radio repair. He was employed at Bird for two years, January to January. Add zip codes to the addresses. (2) Next, he worked from February 198X to May 198X (1½ years) as a supervisor of mechanical maintenance at Borg Corporation, 6024 Southwest 50th Street, Miami, Florida. There he trained assistants and received The Worker of the Year Award. (3) The applicant is presently employed as a service manager for Ace Automation Company, 1102 Broad Way, Pompano, Florida, where he supervises men in all aspects of repair and service. He began work at Ace in June 198X.
2. Revise the following cover letter content to make it more assertive, specific, and persuasive. The applicant is seeking a position as a draftsman with a large, architectural firm.

   Pursuant to your recent ad, I am applying for a job with your company. I don't have much experience but am willing to learn. I will be graduated from college this June, and although I changed my major three times, I will earn a degree in drafting technology.

   I have studied some very pertinent courses and held a number of part-time jobs with local architects learning about materials, office procedures, and on-site supervision. I think I have a pretty good design ability and am a careful draftsman.

   I realize you want a more experienced person, but I would like to talk to you about myself.

3. Compile an employment dossier on yourself. List all of the companies where you have been employed. Note the inclusive dates, complete address, and your position titles. List all of your responsibilities, promotions, achievements, and awards.

   Next, collect your educational records. Obtain the addresses of all schools attended and your final report cards and college transcripts. Figure out your accumulated grade point average. Underline those courses which are job-oriented. Assemble and/or list your extracurricular activities, recognition or reward certificates, and so on.

   Clip newspaper and trade magazine Help Wanted ads in your field. Photocopy the Yellow Pages which list companies which might employ you. Arrange an interview with a career counselor, employment agency representative, or someone in your desired line of work. Take notes and file them in your dossier.

List your three greatest character and work-related strengths.

Include your military records, community organization data, and other pertinent records.

4. Speak to four past employers, faculty members, or other community leaders about supplying letters of reference for you. Inform the prospects about the type of employment you are seeking. Ask each to write a general letter addressed to "To Whom It May Concern" attesting to your qualifications and character.

WRITING OPTIONS

1. Using the sample resume format or any of the three sample resumes in this chapter, write a personal resume for a particular position which you could fill immediately while you continue your college education. Include four references.
2. Using the sample resume format or any of the three sample resumes in this chapter, write a personal resume for a full-time position. Assume that you were graduated last month so that you have completed the courses and earned the requisite degree for your chosen career. Include four references.
3. Write a cover letter to a specific person at a specific company, agency, or institution. Be brief, persuasive, and self-confident.
4. Exchange cover letters with another student and write a negative response to the cover letter. Pretend you are a Director of Personnel and that the applicant either lacks the position qualifications, or fulfills the qualifications but no positions are available at this time. Review the strategies for negative response letters which are discussed in Chapter 2.

4

GRAPHICS

GOOSEMYER by PARKER & WILDER

By permission.

Skills:

After studying this chapter, you should be able to

1. Recognize the purpose and function of bar charts, line graphs, circle graphs, flow charts, organization charts, drawings, maps, and photographs.
2. Distinguish the conventions of incorporating tables and figures.
3. Prepare random, continuation, and formal tables.
4. Prepare a bar chart.
5. Prepare a line graph.
6. Prepare a circle graph.
7. Prepare a flow chart.
8. Prepare an organization chart.
9. Prepare a simple, exploded, or cutaway drawing.
10. Incorporate a map or photograph into a brief report.
11. Devise and prepare the appropriate graphics for a specific writing assignment.

INTRODUCTION

In addition to the visuals discussed in Chapter 1 (short paragraphs, headings, numbers, capitals, underlining, bullets, and so forth), graphic illustrations are characteristic of professional and technical report writing. These are referred to as *tables* and *figures*. Tables include randomly incorporated data, continuation tables, and formal tables. All other graphics are referred to as figures. Figures include bar charts, line graphs, circle graphs, flow charts, organization charts, drawings, maps, and photographs.

Large companies may employ graphic artists to assist writers in graphic illustration, but it is the writer's task to decide on appropriate graphics and to provide, at least, a rough idea of the layout and data. Many simple graphics can be handled by the novice without assistance.

Purpose

Graphic incorporations are never merely decorative. They serve to

- Speed up a reader's comprehension
- Add credibility to the material
- Serve as a method of quick reference
- Add to the attractiveness of the report

Each type of graphic serves a distinct purpose. For instance, a table displays data in vertical columns that would otherwise involve lengthy prose sentences which, in turn, might be difficult to comprehend or to interpret. A bar chart illustrates comparisons of parts while a circle graph shows not only comparisons of parts but also the relationship of each part to the whole. Drawings, maps, and photographs, can show details that words cannot describe. You must consider not only when to use a graphic illustration, but also which type of graphic will best serve your purpose.

General Conventions

Certain conventions are adhered to for all graphic illustrations:

1. All graphics are inserted as close to their textual reference as possible rather than attached on separate pages. Graphics which are included as supplemental or exhibit materials following a report lose their impact.
2. Formal tables are always numbered and titled *above* the data. All other graphics are referred to as figures and are numbered and titled *beneath* the graphic. The numbers and titles may be centered or placed flush left to the margin of the report.
3. If more than one table or more than one figure is used in a report or manual, each is numbered in order of its appearance throughout the material. Arabic numbers are used:

   Table 1
   Table 2
   Figure 1
   Table 3
   Figure 2

   If the report contains numbered chapters, a decimal numbering system is used to indicate both the chapter and the sequential number of the graphic:

   Figure 7.1
   Table 7.1
   Figure 7.2
   Table 7.2

4. A precise *noun-phrase title* is included with each designation for clarification.

   Table 1 Cost Comparison of Transportation Modes
   Table 2 Smoke Detector Ratings
   Table 3 Characteristics of Whales

TABLE 4
MEAN SALT CONCENTRATION
IN VARIOUS SOURCES OF WATER

Figure 1 Cross section of a typical speed bump before and after modification

Figure 2 Proposed Transit Systems Routes

5. Convention 4 indicates a variety of acceptable uses of periods and capital letters. It is important to be consistent, however. If you decide to use a period after the graphic number, do so for all graphic designations throughout your report. You may capitalize an entire title, capitalize initial letters of each word, or capitalize only the initial letter of the first word. Consistency within a report is the key to effectiveness.
6. Titles which require more than one line are usually single spaced. Second and consecutive lines are aligned under the first word of the title, not under the word *table* or *figure*.
7. Graphics are always referred to in the text by an introductory sentence, such as "Figure 7 shows an example of a sample field report investigating a claim for damages:" Graphics are never just "plopped" into a report. Generally a sentence or two of interpretation or emphasis follows the graphic.
8. Graphics are contained within the margins of your report. No labels, headings, portions of drawings, legends, and so forth should extend left or right of your established margins. It is unconventional to place graphics sidewards on a page. Copy machines with reduction capabilities are helpful for wide graphics. Prepare a graphic with the top on the vertical plane of your paper, reduce it with the copy machines, carefully paste it in the appropriate space of your report, and photocopy the entire page. If reduction is not possible, insert the graphic on a separate page with the top of the graphic on the left-hand-side vertical plane.
9. In the case of tables which require more than one full page, begin the second page with the designation, number, and the word *Continued*:

Table 4 Continued

10. Explanatory notes, keys, and legends usually appear within the graphic or beneath the graphic in the left-hand position but above the number and title. See Figure 4.6 for an example of this convention.
11. Identify your source of borrowed graphics in parentheses after the title:

Figure 1 Sample formal proposal (Courtesy of Charles E. Smith, Jr., Robert Heller Associates)

Table 7 Average Yearly Salaries by Sex and Race (Source: Catherine Brown, Discrimination in the Work Place [New York: Silver Press, 1980], p. 8.)

Figure 2 Fire ground injuries by cause ("Fire Ground Injuries in the United States during 1979," Fire Command [December 1980], p. 12. Reprinted by permission.)

TABLES

Tables are visual displays of numerical or nonnumerical data arranged in vertical columns so that the data may be emphasized, compared, or contrasted. Tables may be *informal* (random and continuation) or *formal*. Figures 4.1 and 4.2 illustrate informal random tables in report texts.

The Training Center announces the beginning of a mini-course, "Write It Right—Write It Well," for senior executive secretaries and administrative assistants. Dates, locations, and purposes follow:

Date	Room	Purpose
May 11	102	to review grammar/usage
May 13	102	to review brief report forms
May 18	101	to review graphics
May 20	102	to review manual components
May 25	103	to review formal reports
May 27	101	to critique individual writing

To register, fill out the attached form and forward it to Julie Wood, Training Specialist, Room 608.

Figures 4.1 and 4.2 Sample informal random tables

Regardless of what kind of accident is being reported, certain information must be reported objectively and specifically:

- What the accident is
- When and where the accident occurred
- Who was involved
- What caused the accident
- What were the results of the accident (damage, injury, and costs)
- What has been done to correct the trouble or to treat the insured
- What recommendation or suggestions are given to prevent a recurrence

Information required for the accident report has become so standardized that many companies have designed accident report forms.

Random Tables

Brief lists of figures, dates, personnel, important points, and the like may be displayed in vertical columns for visual clarity and quick reference.

Random Table Conventions

1. Random tables are used only for brief data.
2. Each is introduced by an explanatory sentence.
3. Each is indented 5 to 10 spaces from the left- and right-hand margins of the page.
4. The tabulated data may contain column headings, numbered data, or bullets.
5. A random table does not include a table designation, number, or title.

Continuation Tables

A continuation table is one that contains prose data in a displayed manner. It reads as a continuation of the text and includes the same punctuation marks which would be required if the data were presented in paragraph form. Figure 4.3 illustrates two informal continuation tables in a brief accident report.

Continuation Table Conventions

1. Each is introduced by a sentence followed by a colon if the last introductory word is *not* a verb.
2. The tabular data is indented 5 to 10 spaces from the left and right margins.
3. The continuation table presents an alignment of figures, dates, or other data.
4. The data is punctuated by standard commas, semi-colons, and periods as if the material were presented in paragraph form.

Our insurance policy All State #17B-445-9100K will cover the cost of the fire damage. Repairs and replacements total $390.00. This price includes

$45.00	for carpet replacement (9 sq ft @ $5.00 per ft; Carl's carpets),
20.00	for labor for removing burned carpet and replacing (5 hr @ $4.00 per hr),
15.00	for cleaning solution for wall (15 sq ft @ $1.00 per ft),
35.00	for paint for wall (70 sq ft @ $0.50 per ft)
80.00	for labor for cleaning and repainting wall (10 hr @ $8.00 per hr),
80.00	for fabric for brown leather armchair (8 yd @ $10.00 per yd),
70.00	for labor for reupholstering armchair (7 hr @ $10.00 per hr),
20.00	for new magazine stand from Pier One, and
25.00	for fire extinguisher replenishment.

In order to prevent fires such as this in the future, I recommend that

a) all ashtrays be removed from the waiting rooms,
b) three "No Smoking" signs be placed on each of the end tables, and
c) the magazine rack be placed next to the receptionist's desk to remove it from hazard.

Should you endorse the second recommendation, I will personally obtain the signs from the Davie Fire Department at no cost and place them by Monday, February 4, 198x.

Figure 4.3 Sample informal continuation tables

Formal Tables

Formal tables are used to present statistical information or to categorize and tabulate other written information. Tables 4.1 and 4.2 show two types of formal tables and the conventions adhered to in typing such tables:

TABLE 4.1
TIME/COST FOR AERIAL PHOTOGRAPH SEARCHES

Time Frame	Searches (#)	Time @ 30 min ea (hr)	Cost @ $6.00 per hr wage ($)
Daily	3	1½	9.00
Weekly	15	9	54.00
Monthly	60	30	180.00
Quarterly	180	90	540.00
Yearly	720	360	2160.00

Table 4.2 Troubleshooting Chart for Heath Kits

Difficulty	Possible Cause
Receiver section dead	Check V1, V3, V4, V7, and V8
	Wiring error
	Faulty speaker
	Faulty receiver crystal
	Crystal oscillator coil mistuned
Receiver section weak	Check V1, V2, and V3
	Antenna, RF or IF coils mistuned
	Faulty antenna or connecting cable
Transmitter appears dead	Check V5 and V6
	Wiring error
	Recheck oscillator, driver, and final tank coil tuning
	Dummy load shorted on open

Either the centered or flush-left title is conventional.

Formal Table Conventions

1. Formal tables contain horizontal lines from margin to margin above the title (optional), below the title, below the body of the table, and between the column headings (the boxhead) and the body of the table.
2. The boxhead of vertical column headings indicates the body figure symbols in parentheses [*i.e.* ($), (rpm), (hr), (ft)].
3. Formal tables are not closed on the sides.
4. The columns are always vertical. The first body column is called the *stub*.
5. In modern practice the body does not contain leaders (spaced periods to aid the eye in following data from column to column).

FIGURES

As previously mentioned, all graphics except tables are referred to as figures both in textual reference and in titling.

Bar Charts

Bar charts are used to show differences in quantity visually and instantaneously. They are frequently made from a statistical table source. They are commonly used to show quantities of the same item at different times, quantities of different items for the same time period, or quantities of the different parts of an item that make up the whole.

Bars may be plotted vertically or horizontally. It is conventional to plot bars representing monetary units into vertical graphs.

Use graph paper to plot your chart. The scale you select is critical to the success of your chart. Do not include grids which will not contain some portion of a bar. All grids must be scaled to equal increments, such as 0, 1, 2, 3, or 0, 5, 10, 15, but not 0, 5, 7, 12.

If the order of the bar placement is not controlled by a sequential factor, place the longer bars to the bottom of a horizontal chart or to the right of a vertical chart. This placement avoids a top heavy or one-sided chart. Figure 4.4 shows a horizontal bar chart where placement is not a factor; therefore, the longest bars have been placed at the bottom. Figure 4.5 shows a vertical bar chart where the bars must be placed in sequential order; however, in this case the largest bars naturally fall to the right side of the chart. A bar chart may illustrate up to three or four comparative bars in groupings. Figure 4.6 illustrates a multiple bar graph.

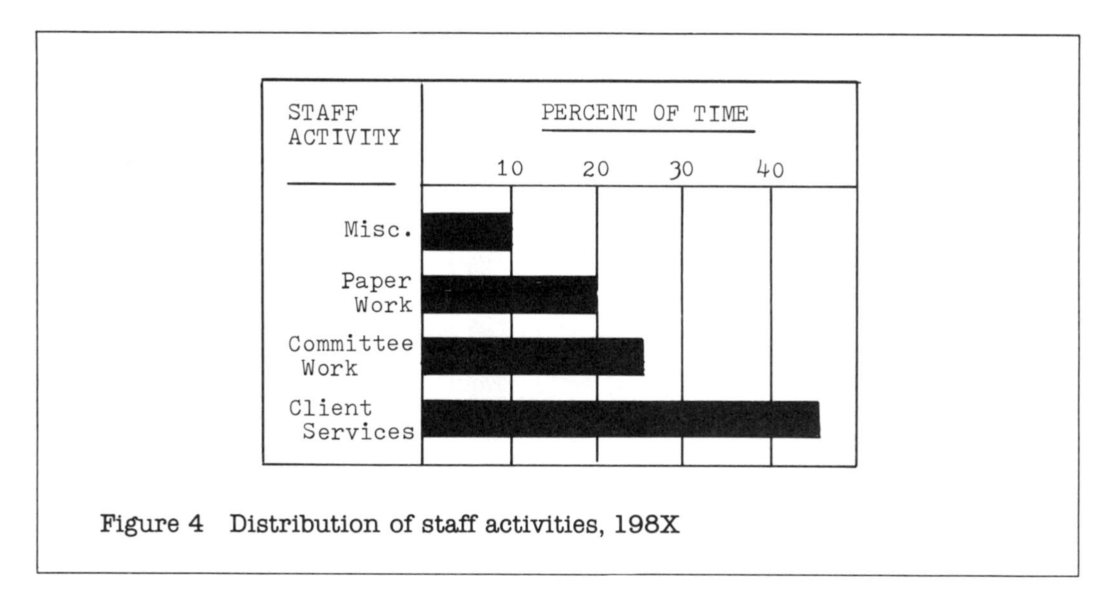

Figure 4.4 Typical horizontal bar chart

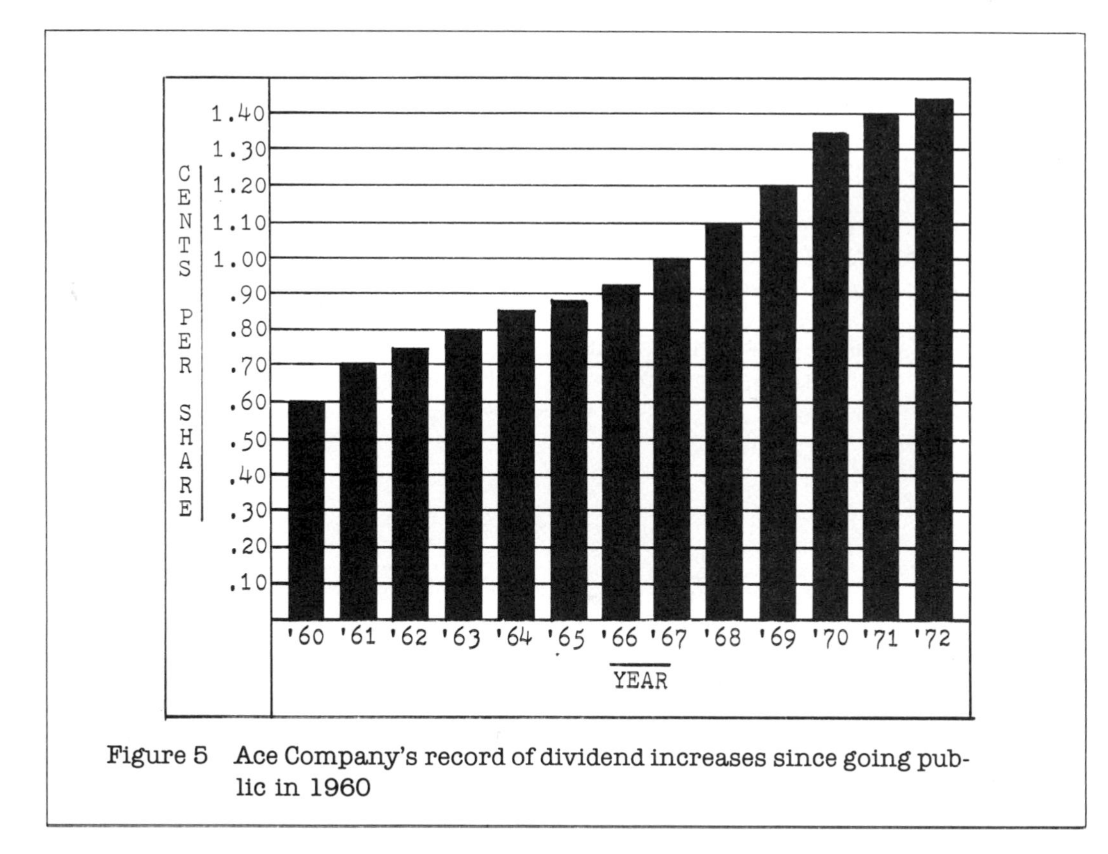

Figure 4.5 Typical vertical bar chart

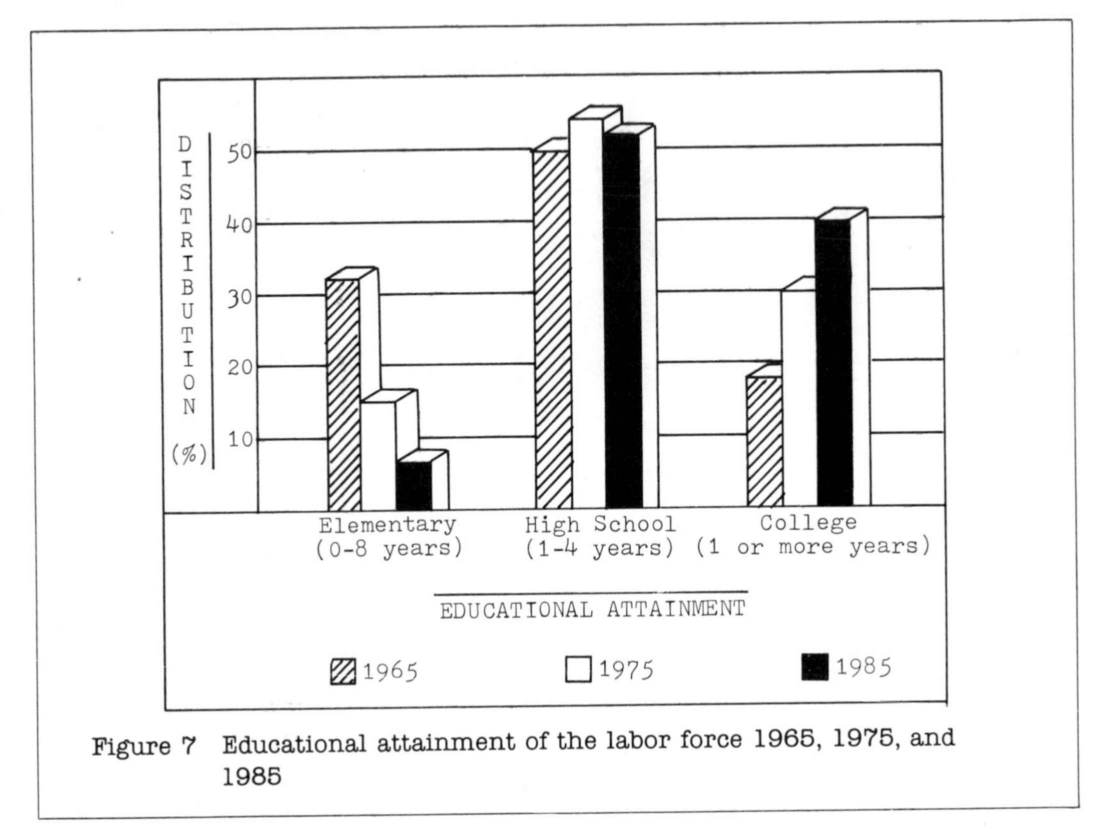

Figure 7 Educational attainment of the labor force 1965, 1975, and 1985

Figure 4.6 Typical multiple bar chart

Shadings and hatchings add interest and emphasis to multiple bar representations. Dimension adds depth, volume, and body.

Bar Chart Conventions

1. All of the headings, legends, and so forth are contained within a box.
2. Bars of even width should be evenly spread.
3. Partial cut-off lines separate headings from grid notations.
4. The horizontal or vertical grids are usually shown, but not both in a single chart.
5. Grid notations are centered on the grid lines.
6. All headings read horizontally, if possible.

Line Graphs

Line graphs, or curves, are used to show changes in two values. Most commonly, they show a change or trend over a given period or a performance with a variable factor.

Like bar charts they are usually boxed. The curve is always plotted from left to right. Both horizontal and vertical grid lines of equal increments are drawn in or indicated by tick marks. The lower left-hand intersection of the grids is the key point from which both incremental grids begin. The horizontal axis contains the grid notations for the static values (years, months, speeds, or other set units). The vertical axis shows incremental grids for quantity or amount factors. The vertical axis is graduated into equal proportions from the least amount at the bottom to the greatest amount at the top.

Figure 4.7 shows a simple sales curve over a seven-month period. Figure 4.8 illustrates a multiple line graph. Notice that the zero point of the graph is not included because to do so would result in an extraneous, unused grid. Lines which serve both as outline and grid are not labeled; it is obvious to the reader that the increments are equal and sequential.

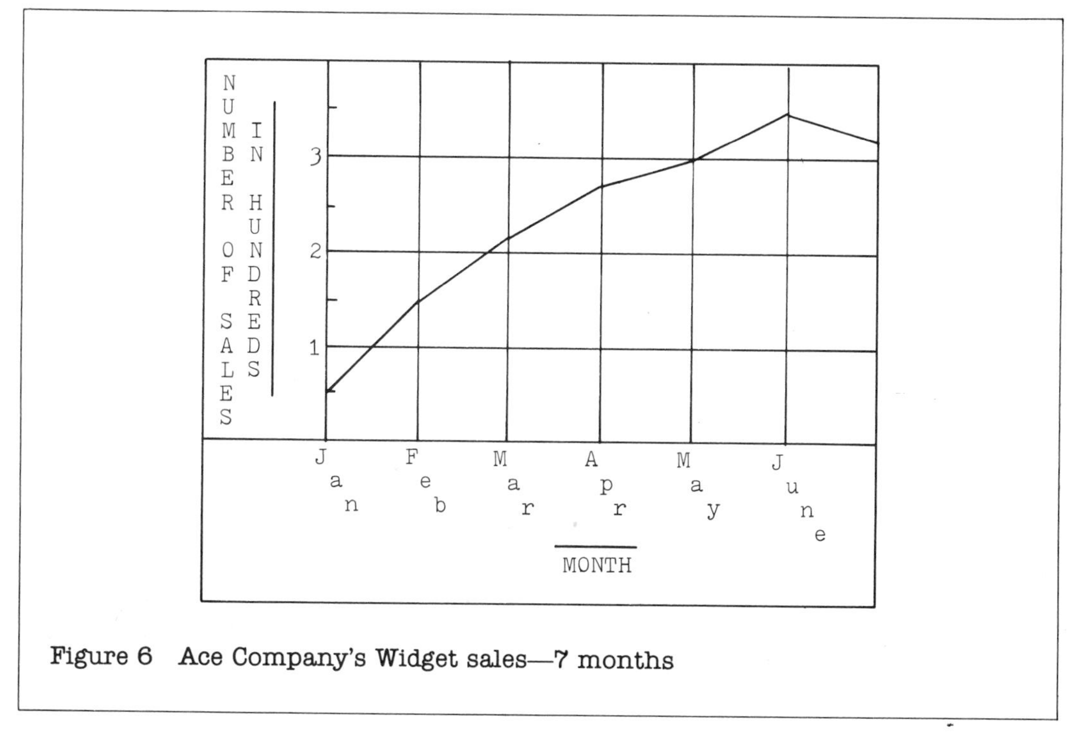

Figure 6 Ace Company's Widget sales—7 months

Figure 4.7 Typical line graph

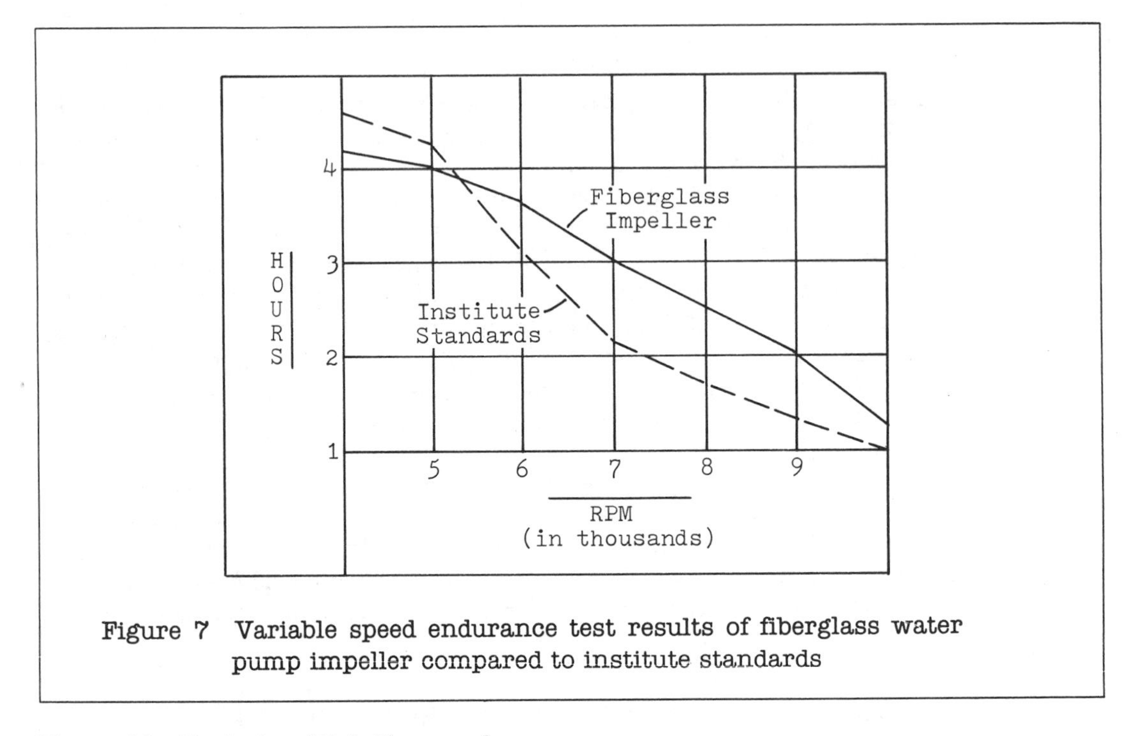

Figure 7 Variable speed endurance test results of fiberglass water pump impeller compared to institute standards

Figure 4.8 Typical multiple line graph

Line Graph Conventions

1. All of the information is usually boxed.
2. Curves are always plotted from left to right.
3. Both horizontal and vertical grids are drawn in or indicated by tick marks.
4. Major headings are capitalized; subheadings or grid notations use initial-letter capitals only.
5. Grid notations are centered on the line or tick marks.
6. Usually partial cut-off lines separate headings from grid notations.
7. All headings read horizontally, if possible.

Circle Graphs

Circle graphs, also known as pie charts, circle charts, pie diagrams, or sector charts, are used to compare the relative proportions of various factors to each other and to the whole. The circle represents 100 percent while the segments indicate the proportionate percentage of each factor to the whole. They are popular for illustrating financial information or survey responses.

Because a circle contains 360 degrees, you must first convert your real data to percentages of the whole and then calculate the number of degrees needed to represent each segment or wedge. Use the following format to calculate the number of degrees for each segment:

	Item	Raw Data	Frac-tion	4 place decimal	3 place decimal	Percent	Percent rounded	Degrees % × 360°	Degrees rounded
Ex:	Rent	$400	$\frac{400}{2300}$	.1739	.174	17.4%	17%	61.2°	61°
	____	____	____	____	____	____	____	____	____
	____	____	____	____	____	____	____	____	____
	____	____	____	____	____	____	____	____	____
		TOTAL:	$2300		TOTAL:	100%		TOTAL:	360°

A compass or circle template and protractor are needed to draw the circle and to divide it into segments. As a rule of thumb use a 3-in. diameter circle on standard 8½ by 11 in. paper. This will make your circle large enough for emphasis yet small enough to allow space for labeling the wedges. Think of your circle as a clock, and plot your largest wedge in the upper right-hand quadrant from the 12 o'clock position. The wedges then decrease in size clockwise with proportionately smaller wedges or segments.

A circle graph is effective without shading although you may use shading and hatching to add interest and to differentiate further each segment of the circle. Labels and the percentage are generally shown outside of each wedge to avoid crowding. Center each label on the radius of each wedge or use a tag line to aid the eye. All labels must be contained within the left- and right-hand margins, and all are typed on a horizontal plane. Figure 4.9 shows a simple circle graph with tag lines to center the segment labels. Figure 4.10 shows a circle graph with separated segments.

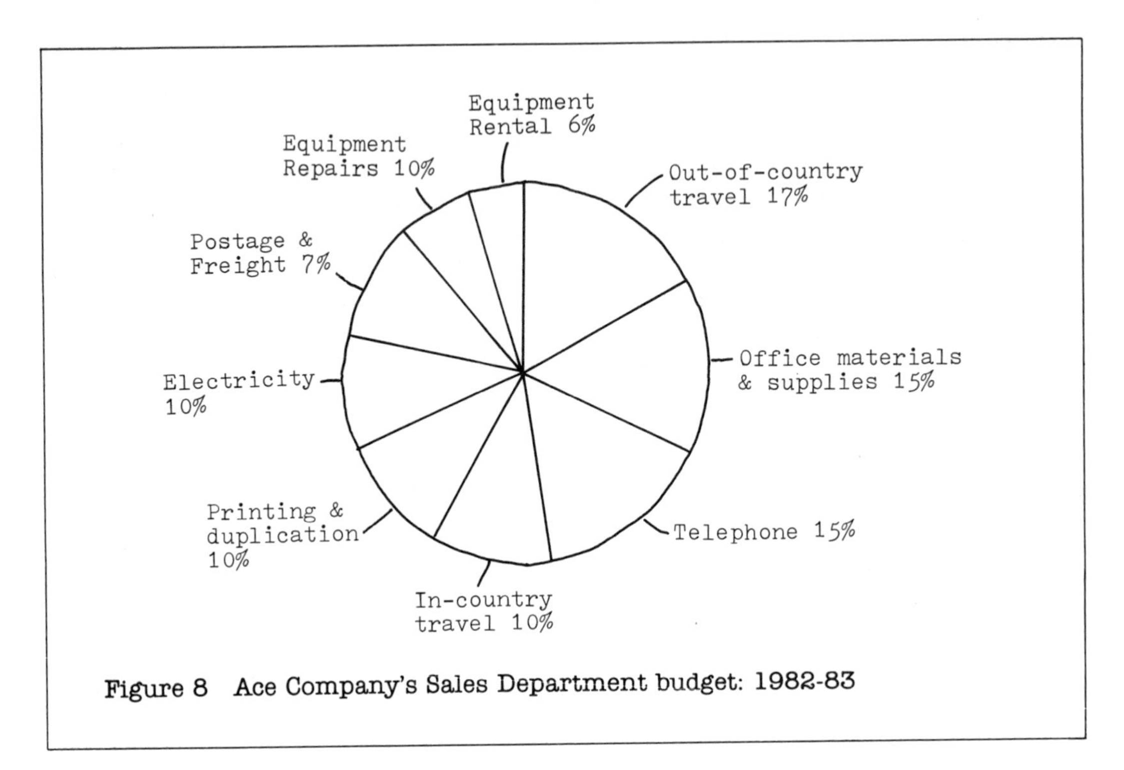

Figure 8 Ace Company's Sales Department budget: 1982-83

Figure 4.9 Typical circle graph

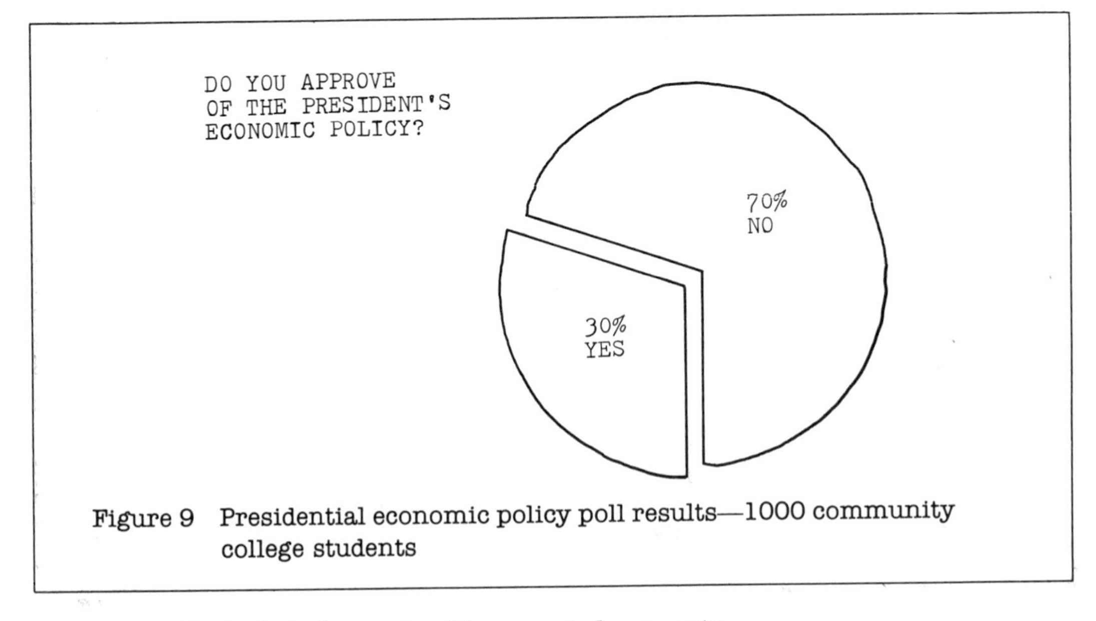

Figure 9 Presidential economic policy poll results—1000 community college students

Figure 4.10 Typical circle graph with separated segments

Circle Graph Conventions

1. Circles are usually no larger than 3-in. in diameter.
2. The largest segment is placed in the upper right-hand quadrant with the segments decreasing in size clockwise.
3. Segment labels and the represented percent are usually typed outside of the segments, centered on the segment radius.

Computer programs now available can construct a number of graphics which, in turn, can be printed by a word processor or a multi-pen plotter. Although the graphic printout on the plotter fills an 8½ by 11 in. sheet of paper, photoreduction of the hardcopy can make the graphic a suitable size for incorporation into your report. Figures 4.11, 4.12, and 4.13 show computer printouts of a bar chart, a line graph, and a circle graph respectively, each printed by a TRS-80 Model III microcomputer with a Tandy-Graph Multi-Pen Plotter:

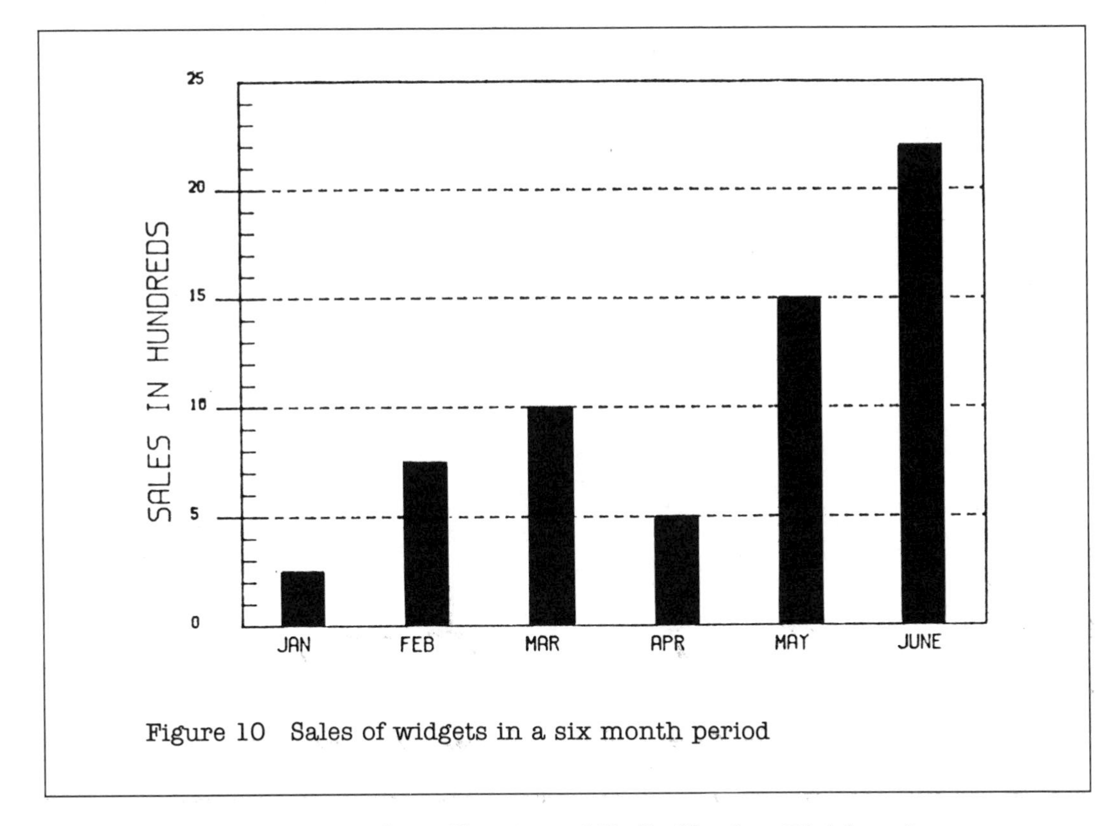

Figure 4.11 ***Computer bar chart (Courtesy of Radio Shack, a Division of Tandy Corporation, Computer Center, Hollywood, Florida)***

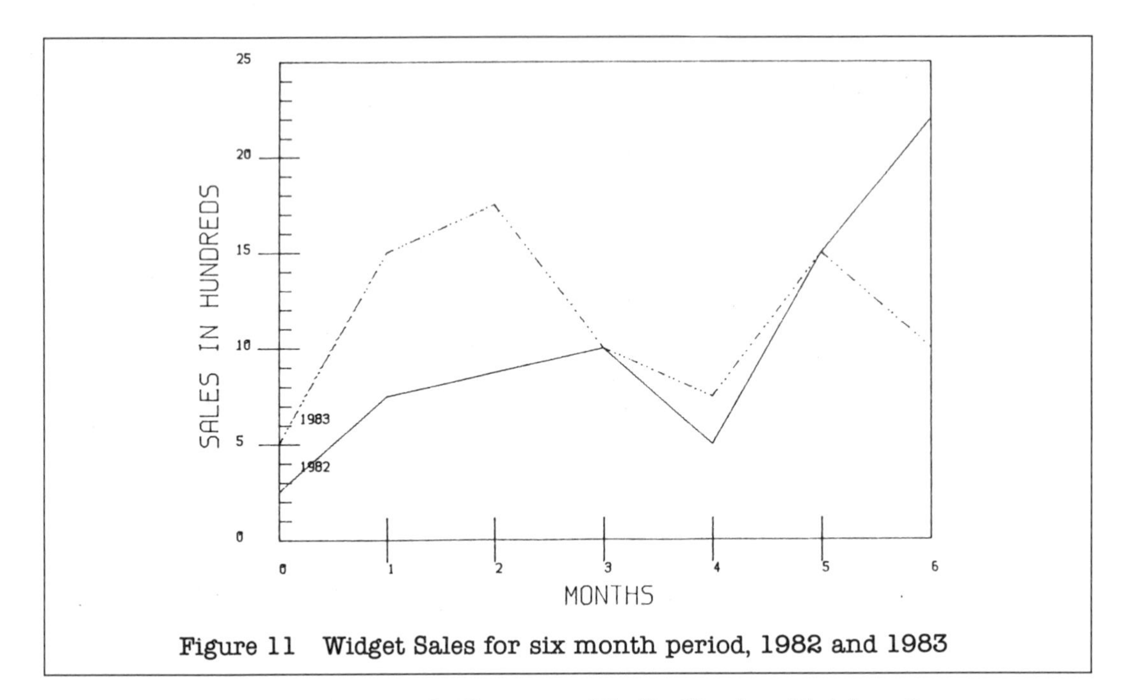

Figure 4.12 Computer line graph (Courtesy of Radio Shack, a Division of Tandy Corporation, Computer Center, Hollywood, Florida)

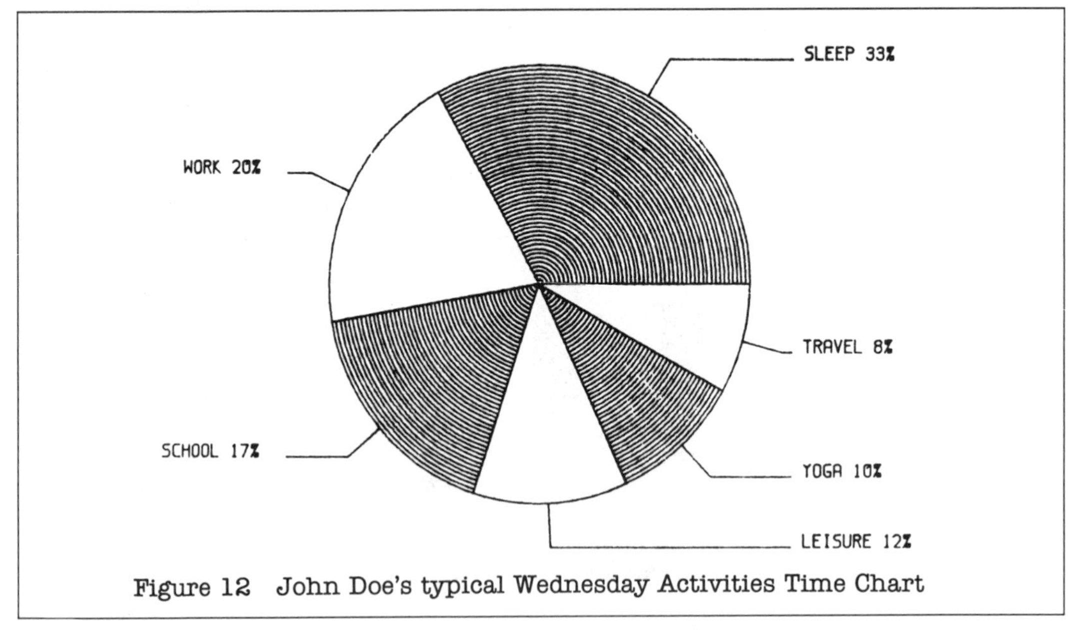

Figure 4.13 Computer circle graph (Courtesy of Radio Shack, a Division of Tandy Corporation, Computer Center, Fort Lauderdale, Florida)

The multigraph printer can print up to 20 bars singly or in multiple-bar groupings with six variable shadings. Nine different lines, solid or dotted, are possible for multi-line graphs. The Tandy-graph can plot up to 15 wedges, with or without shadings. Notice that the computer does not provide outline boxes for the bars and curves, nor does it print partial cut-offs between the headings and grid notations. These conventions may be added if desired.

Flow Charts

A flow chart is used to show pictorially how a series of activities, procedures, operations, events, or other factors are related to each other. It shows the sequence, cycle, or flow of the factors and how they are connected in a series of steps from beginning to end. The information is qualitative rather than quantitative as in bar charts, line, and circle graphs.

The components of a flow chart may be diagrammed in horizontal, vertical, or circular directions. They condense long and detailed procedures into a visual chart for easy comprehension and reference. Computer programmers use templates which contain a variety of shapes symbolizing various activities, such as

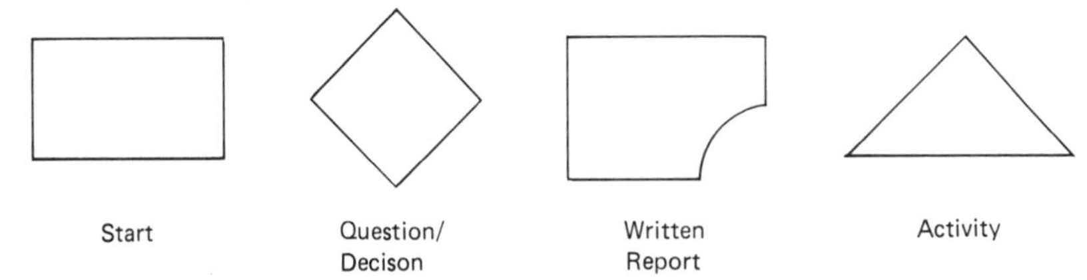

Usually boxed steps are arranged in sequence and connected by arrows to show the flow. Flow charts may be simple or pictorial. Figures 4.14 and 4.15 illustrate simple and pictorial flow charts:

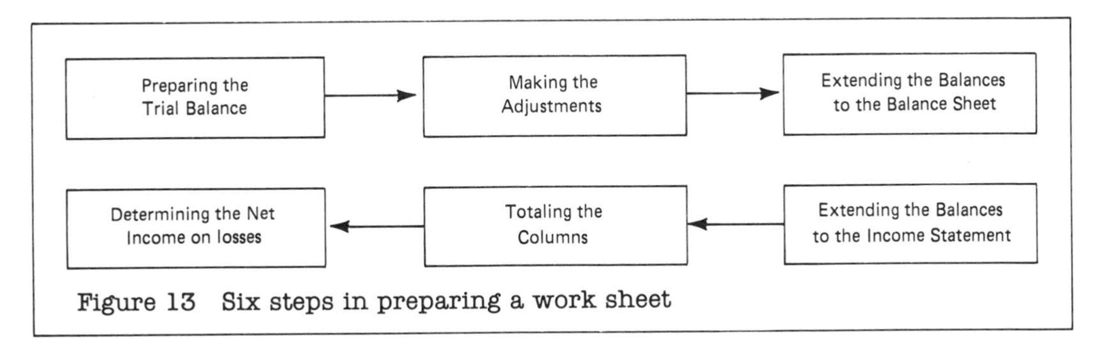

Figure 4.14 Typical flow chart

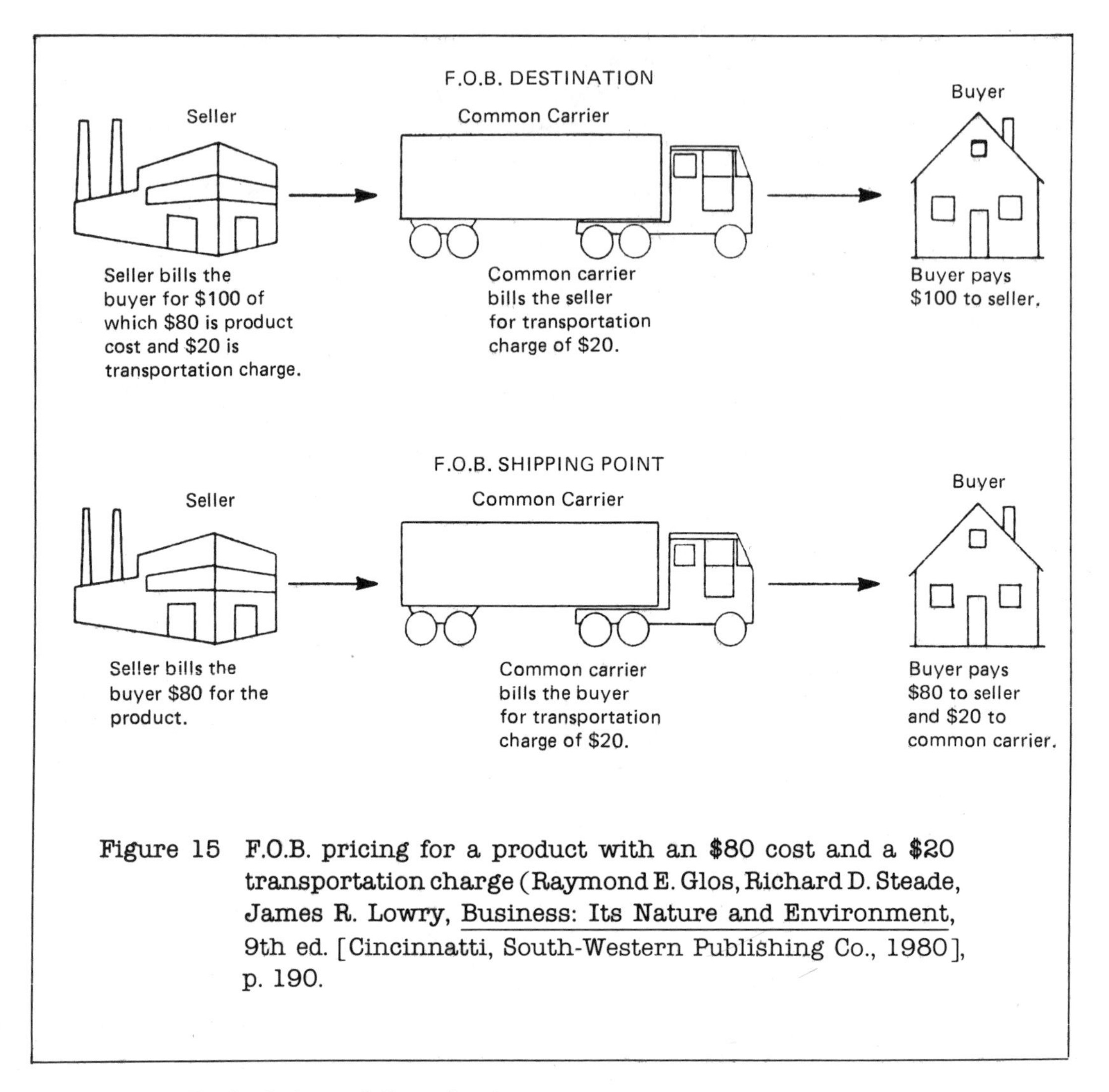

Figure 4.15 Typical pictoral flow chart

Flow Chart Conventions

1. Flow charts employ squares, boxes, triangles, circles, diamonds, and other shapes to enclose each step.
2. Steps naming major activities are typed within the shapes.
3. The flow of steps may be horizontal, vertical, circular, or a combination of directions.
4. Lines or arrows connect the shapes to show the flow of activities.

Organization Charts

Like flow charts, organization charts show quantitative, rather than qualitative material. The organization chart is used to show the relationship of the organization's staff positions, units, or functions to each other.

A staff organization chart shows the chain of command of the staff positions, such as President, Vice-presidents, Directors, Controller, Personnel Director, Salespersons, and so forth. A unit organization chart depicts the relationships among such units as Public Relations Department, Research Division, Finance Office, Personnel Section, and so forth.

A function chart shows the span of control of such functions as Planning, Marketing, Production, Data Processing, and so forth. These three aspects—staff, unit, functions—should not be mixed together in the same chart.

Either a horizontal or a vertical emphasis can be imparted to an organizational chart by the layout:

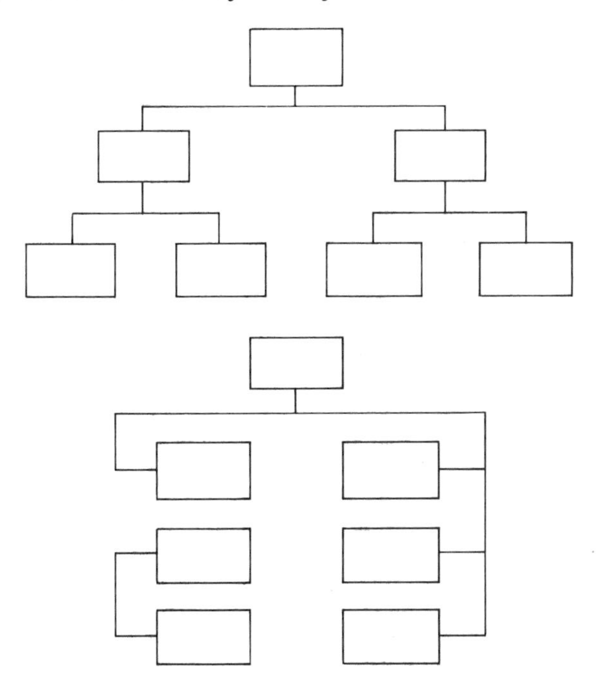

Figure 4.16 shows a staff organization chart for a manufacturing company.

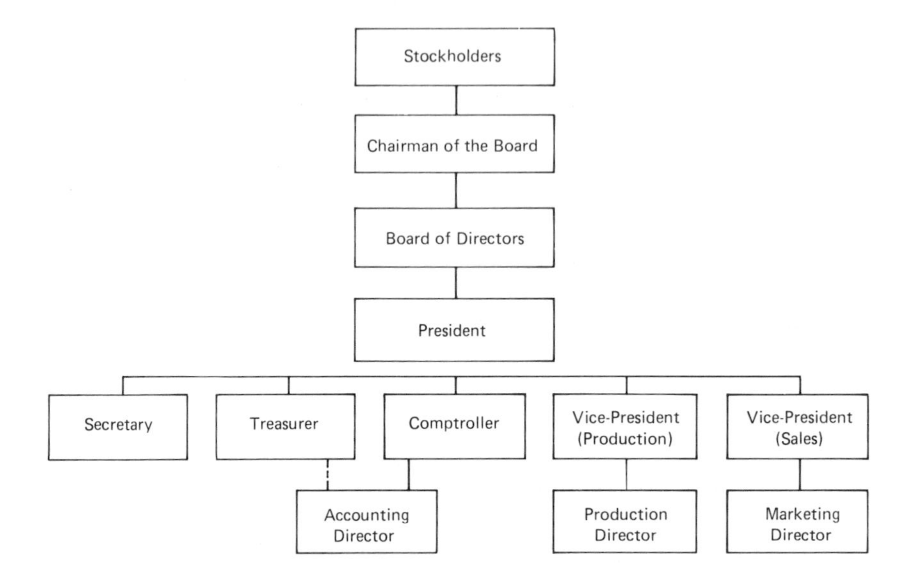

Figure 4.16 Typical organization chart

Organization Chart Conventions

1. Organization charts usually use rectangular boxes to enclose staff, unit, or function titles.
2. Solid lines represent the relationship, line of authority, or chain of command between the units; a broken line may represent an open line of communication for reporting out of the chain of authority.
3. A horizontal or vertical emphasis may be suggested by the chart layout.

Drawings

A variety of simple drawings may be executed by the novice. A simple line drawing is often clearer than a photograph because the former can give emphasis to the major parts. Diagrams of procedures can be useful for clarifying instructions. Exploded-view illustrations show the proper sequence in which parts fit together. A cutaway drawing can show the internal parts of a mechanism or piece of equipment. Electricians and electronic technicians use schematics and wiring diagrams to illustrate concepts.

Drawings should be uncluttered, properly ruled, and carefully labeled. Figures 4.17 through 4.21 show a variety of simple drawings which can clarify descriptions, instructions, definitions, and process analyses:

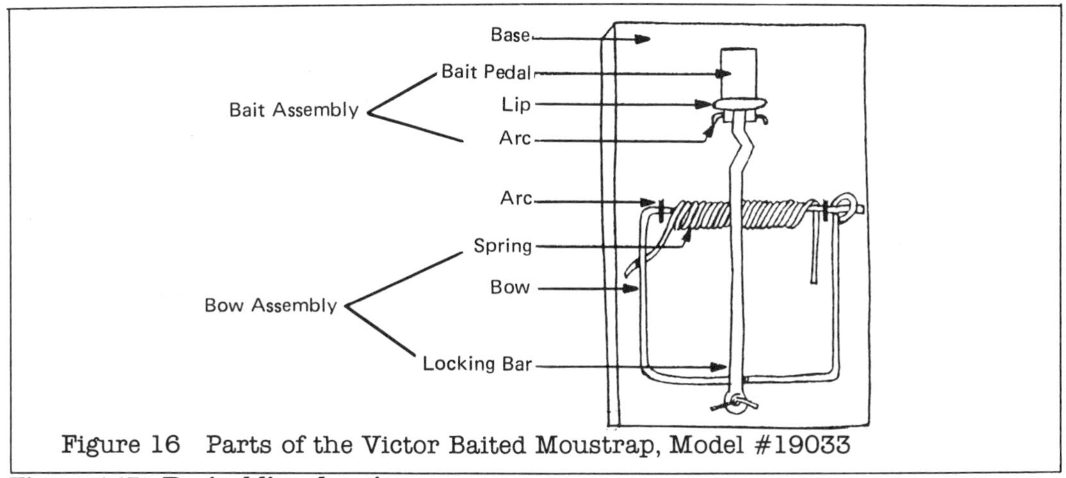

Figure 16 Parts of the Victor Baited Moustrap, Model #19033

Figure 4.17 Typical line drawing

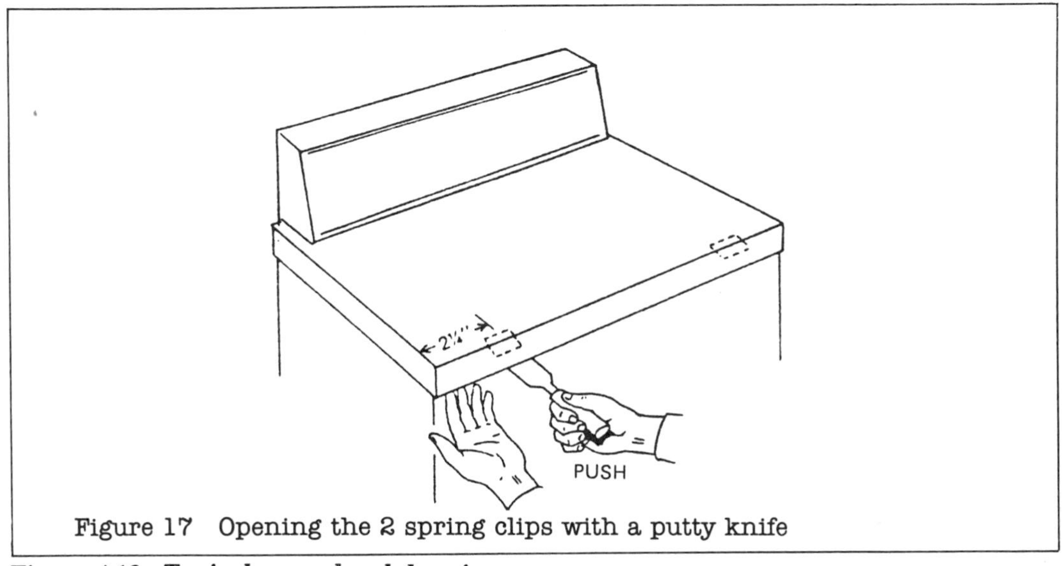

Figure 17 Opening the 2 spring clips with a putty knife

Figure 4.18 Typical procedural drawing

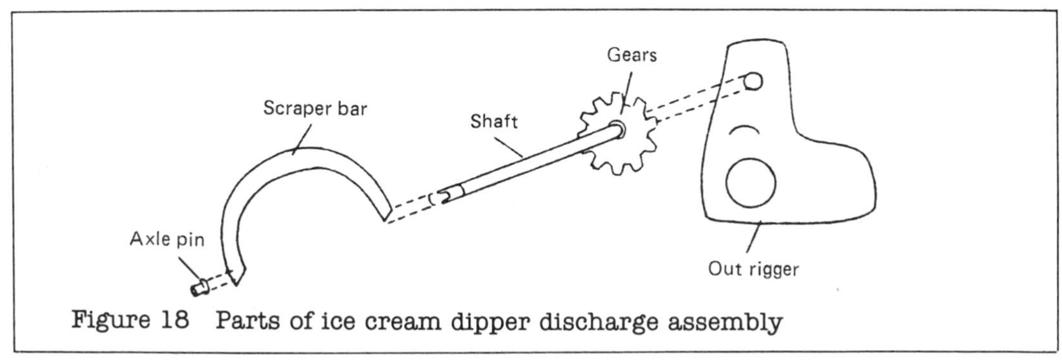

Figure 18 Parts of ice cream dipper discharge assembly

Figure 4.19 Typical exploded view illustration

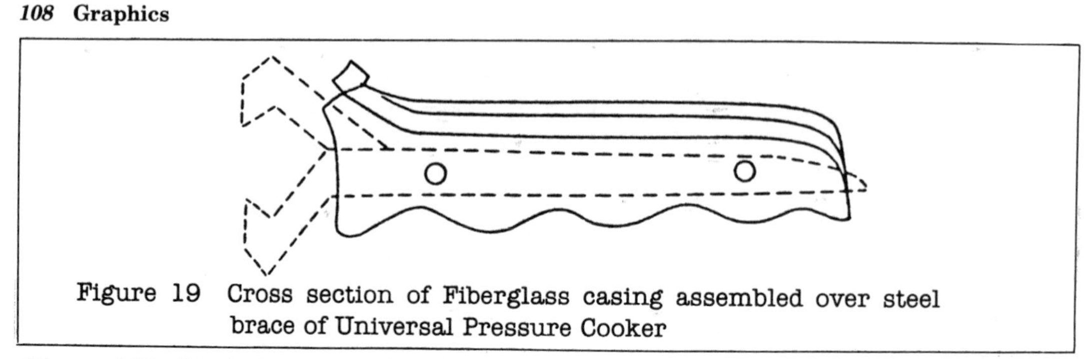
Figure 19 Cross section of Fiberglass casing assembled over steel brace of Universal Pressure Cooker

Figure 4.20 Typical cutaway drawing

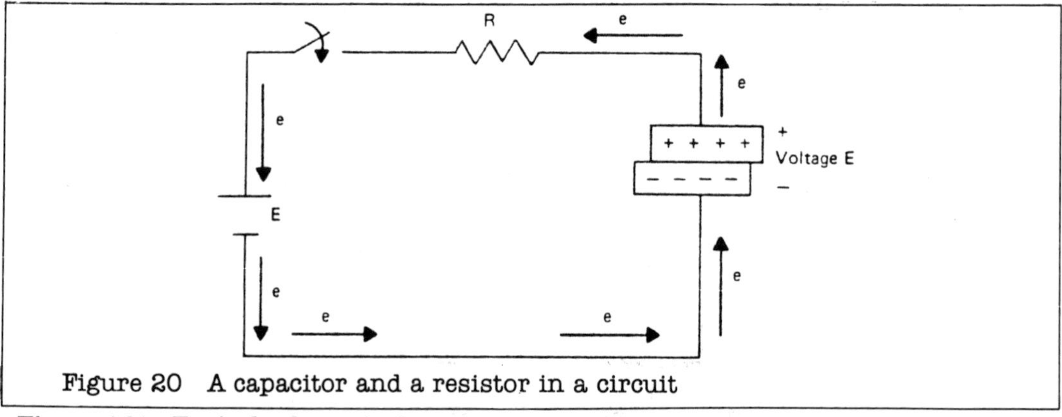

Figure 20 A capacitor and a resistor in a circuit

Figure 4.21 Typical schematic drawing

Drawing Conventions

1. All drawings should be simple and uncluttered.
2. All labels of pertinent parts are typed.
3. Dotted lines are shown to indicate relationships of internal or connecting parts.

Maps and Photographs

Maps may be hand drawn or photocopied to illustrate such elements as specific locations, geographical factors, and routes. Photographs are useful to illustrate complex mechanisms and equipment or to provide evidence of such important details as damage or injuries to persons, automobiles, and buildings. Photographs show actual appearance.

Both maps and photographs must be used with care. Both may show attention-diverting and insignificant detail which detract from the emphasis you wish to impart. Figures 4.22, 4.23, and 4.24 show maps and photographs used in a variety of reports:

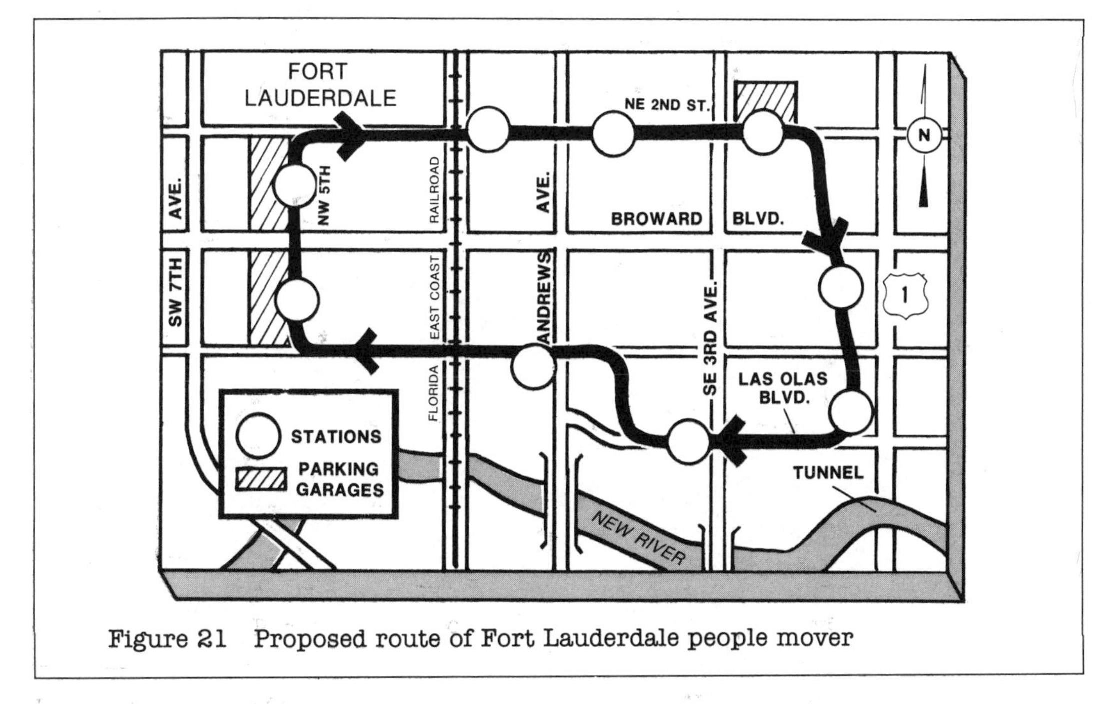

Figure 21 Proposed route of Fort Lauderdale people mover

Figure 4.22 Typical large-scale map (Fort Lauderdale News/Sun Sentinel map by Tom Alston)

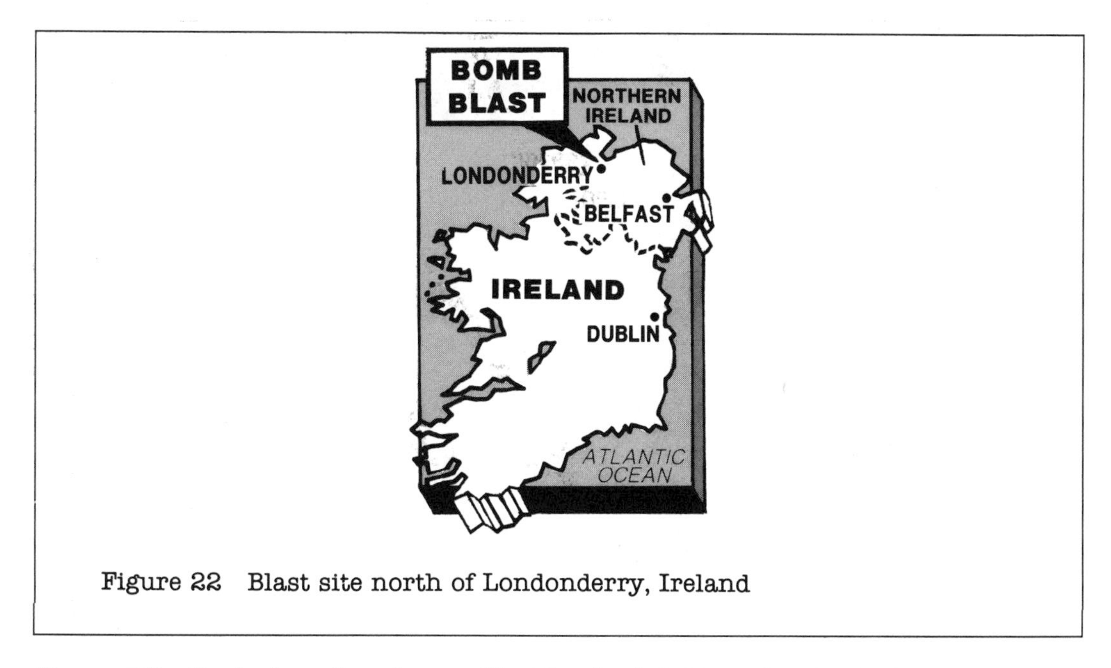

Figure 22 Blast site north of Londonderry, Ireland

Figure 4.23 Typical small-scale map (Fort Lauderdale News/Sun Sentinel map by Keith Robinson)

Figure 23 Front end damage to automobile

Figure 4.24 Cropped photograph emphasizing detail (Courtesy of George Chillag)

EXERCISES

1. Devise a formal table to present the following data obtained from a college study on sex discrimination:

   Of 215 faculty members 79 or 36.7 percent are women

   Of 36 administrators 3 or 8.3 percent are women

   Of 9 counselors 3 or 33.3 percent are women

   Of 7 librarians 5 or 71 percent are women

Of 18 standing committee chairpersons 2 or 11.1 percent are women

Of 207 committee members 52 or 25 percent are women

Number and title your graphic.

2. Figure out how many hours you spend on a typical Wednesday in each of the following activities:

Travel	Grooming
Study/School	Leisure
Work	Sleep
Meals	

Present this information in a horizontal bar chart. Number and title your graphic.

3. Devise a line graph to show the population of an imaginary town over a thirty-year span:

1955	20 thousand
1960	50 thousand
1965	26 thousand
1970	65 thousand
1975	80 thousand
1980	115 thousand
1985	70 thousand

Number and title your graphic.

4. Devise a flow chart to represent visually the basic steps in ordering and receiving materials in manufacturing.

Step 1	Establishing specifications
Step 2	Recognizing needs and activating purchase
Step 3	Selecting the vendor
Step 4	Preparing the purchase order
Step 5	Receiving the goods
Step 6	Evaluating the vendor's performance
Step 7	Updating the vendor's ratings

This is tricky because Step 7 affects Step 3. Show the relationship. Number and title your graphic.

5. Devise an organizational chart to depict the chain of command of data processing positions in a typical business. A data processing

manager oversees all other positions. A systems manager, a programming manager, and an operations manager report to the data processing manager. Under the systems manager's authority are two systems analysts. Two programmers report directly to the programming manager. The operations manager oversees the work of two computer operators and one word processor. Number and title your graphic.

6. Select a small mechanism, hand tool, or kitchen implement and construct a simple or exploded drawing of it. Label all parts including nuts, bolts, rivets, handles, and so forth. Number and title your graphic.
7. Obtain a bus route map or a clear photograph which illustrates a procedure in your field of study. Write a brief paragraph into which it is appropriate to insert the graphic. Introduce, number, title, and comment on the graphic.

WRITING OPTION

You are a marketing specialist at National Motors Corporation. Write a memorandum to the Vice-president of Marketing concerning data on the new-model Econo automobile. Your purpose is to suggest some facts which should be stressed in the new-model Econo sales promotion materials. Use an informal table, a formal table, circle diagram, and a bar chart or curve in the text of your memorandum. Number and title your figures.

The Econo was the most popular car compared to four other competitive mid-sized cars on the market last year. Of total sales 30 percent of buyers chose Econo, 23 percent selected Car A, 20 percent chose car B, 17 percent chose car C, and 10 percent chose car D.

The Econo has doubled its fuel efficiency in five years. Four years ago the Econo was rated at 10.5 miles per gallon (mpg); three years ago the Econo averaged 14 mpg; two years ago the fuel efficiency increased to 17 mpg; a year ago it increased to 20.5 mpg, and this year it has a 21 mpg rating.

Despite base price increases the new Econo is as economical to own and to operate as it was four years ago due to fuel economy. Four years ago the base sticker price was $5801. A 20 percent down payment of $1160 resulted in a $4641 balance to finance. The monthly cost of financing for 48 months was $125 at 13.2 percent. The fuel expense per month for 1250 miles of driving ($15,000 miles annually) at $1.30 per gallon was $116 at 14 mpg.

The cost per month over 48 months (finance charge plus fuel cost) was $241. The new Econo has a base sticker price of $7301. A 20 percent down payment of $1460 leaves $5841 to finance. The monthly cost of financing over 48 months is $170 at 17.7 percent. But fuel expense is reduced to $77 at 21 mpg. Therefore, the cost of owning a new Econo over 48 months (finance charge plus fuel cost) is only $247.

5

BRIEF REPORTS

Skills:

After studying this chapter, you should be able to

1. Appreciate the diversity and purpose of a wide variety of standard reports.
2. List and define common types of brief reports.
3. Devise appropriate formats for conveying report information.
4. Understand that appropriate language, organization, headings, visuals, and graphics enhance brief reports.
5. Write an incident report.
6. Write a periodic or progress report.
7. Write a feasibility report.
8. Write a laboratory or test report.
9. Write a field report.

INTRODUCTION

In the working world the writing of informal and formal reports is an everyday task. Although your particular job description may not detail your writing responsibilities, helping to keep the written record is everyone's duty.

Reports record the events and progress of work. They assist management in making decisions, or they may become working documents to help employees carry on a program.

Types of Reports

Many companies print forms for the types of reports which are required most often, such as accident report forms, travel expense forms, work logs, leave request forms, and so forth. Figures 5.1, 5.2, and 5.3 show typical printed report forms.

Because the information required for other reports is so variable, you must learn to devise your own written accounts. It is helpful to classify reports. Here is a list of typical reports required in business and industry:

Tech/Engr # ___ Shift ___ Date MONTH ___ DAY ___ YEAR ___

Corrective Maint ___ Preventative Maint ___ Other ___

Time on Job HRS ___ MIN ___ Log Item # ___

Work Location ___ Item Continued ___

Complaint/ Work to be done ___

Work Performed ___

Over

Figure 5.1 Typical preprinted work log report

- Expense report (an itemized accounting of all expenses incurred while performing duties, such as promoting sales, attending conferences, or inspecting field progress).
- Feasibility report (sometimes called an *analytical report*; the written evaluation of data to determine the practicality of future products, expansion programs, new equipment or services).
- Field report (the written analysis of data to determine appropriate action, such as estimating real estate value, determining service costs, or establishing claims for damage).
- Incident report (sometimes called an *investigative report*; the written record of an unforeseen occurrence, such as accidents, machine breakdowns, delivery delays, cost overruns, production slowdowns, or personnel problems).
- Laboratory report (sometimes called a *test report*; the written results of laboratory or testing experimentation).
- Periodic or Progress report (the written information on the status of a project).

Reports may vary from a few sentences in a memorandum or on a pre-printed form to several volumes in bound folders, but all share common elements of organization, format, and presentation. By reviewing these elements in five common types of reports, you should attain the skills which apply to writing short reports. Chapter 6 will discuss longer reports and their special features.

RYDER TRUCK RENTAL, INC.

TRAVEL EXPENSE REPORT

VENDOR NUMBER	REFERENCE NUMBER	LOC. CODE	MO. REC.

VEHICLE NUMBER	ACCOUNT NUMBER	AMOUNT
W/E DATE		TOTAL AMOUNT

NAME

TITLE

WEEK ENDING

AIR TRAVEL CHARGED TO RYDER (FOR H.Q. AND REGION MANAGER USE ONLY)		
TICKET NUMBER (last 3 digits)	TICKET DATE	TICKET AMOUNT
		$

BUSINESS PURPOSE OF EACH TRIP		
DATE OF		EXPLANATION (IF NOT CHECKED BELOW)
DEPART.	RETURN	
		☐ VISIT TO COMPANY LOCATION ☐ SALES CALL
		☐ VISIT TO COMPANY LOCATION ☐ SALES CALL
		☐ VISIT TO COMPANY LOCATION ☐ SALES CALL

		SUN	MON	TUES	WED	THURS	FRI	SAT	TOTAL
	DATE								
DAILY ITINERARY	FROM								
	TO								
	TO								
AIR TRAVEL PAID BY EMPLOYEE									
CAR RENTAL*									
PERSONAL CAR EXPENSE (DETAIL ON REVERSE SIDE)									
ROOM*									
MEALS PLUS TIPS**									
COMPANY CAR EXPENSE (DETAIL ON REVERSE SIDE)*									
MISCELLANEOUS (DETAIL ON REVERSE SIDE)									
ENTERTAINMENT (DETAIL ON REVERSE SIDE)**									
TOTAL									

ACCOUNTING FOR ADVANCES		
ADVANCE RECEIVED		
EXPENSES THIS VOUCHER		
BALANCE DUE COMPANY		
OR BALANCE DUE TRAVELER		

* ATTACH RECEIPTS
** INCLUDE ENTERTAINMENT MEALS ON REVERSE

IN ADDITION TO THE REQUIRED RECEIPTS AS SPECIFIED ABOVE, RECEIPTS FOR EACH EXPENDITURE OF $25.00 OR MORE MUST BE ATTACHED.

TRAVELER'S SIGNATURE

APPROVED BY (SIGNATURE)

6-22 (11/80) SIDE ONE 10395 Litho By R In U.S.A.

Figure 5.2 Typical preprinted travel expense report (Courtesy of Ryder Truck Rental, Inc.)

PERSONAL CAR EXPENSE

DATE	SUN	MON	TUES	WED	THURS	FRI	SAT	TOTAL
ODOMETER READING-ENDING								
ODOMETER READING – BEGINNING								
MILEAGE								
LESS PERSONAL MILEAGE								
NET COMPANY MILEAGE								
AMOUNT AT ________¢ PER MILE								

DATE	AMOUNT	VEHICLE NO.	COMPANY CAR EXPENSE (DETAIL BELOW)

DATE	AMOUNT	MISCELLANEOUS (DETAIL BELOW)

ENTERTAINMENT EXPENSE

DATE	AMOUNT	TYPE OF ENTERTAIN-MENT	PLACE NAME, ADDRESS, OR LOCATION	BUSINESS RELATIONSHIP OF INDIVIDUALS OR GROUP ENTERTAINED (Give Name, Title, Etc., Include Names of Co. Employees.)	BUSINESS PURPOSE Date, Duration, Place, and Nature of Associated Business Discussion or how otherwise related to active conduct of the business.

SIDE TWO

Figure 5.2 Continued

REQUEST FOR LEAVE OF ABSENCE*
FIRE DEPARTMENT

NAME ______________________
Last First

DATE ______________________

*1 Leave requests for vacation and anticipated leaves must be submitted two weeks prior to requested starting date.

*2 Leave requests for sick leave and emergency leave must be submitted by 9:00 A.M. the day before your shift works.

____
Signature

LEAVE CODES

V — Vacation
S — Sick with pay
FS — Family Sick with pay
I — Job Injury with pay
Z — Sick without pay
W — Personal Absence w/o pay
A — Absence w/o leave
PG — Maternity Leave w/o pay
FL — Funeral Leave with pay (next of kin)
M — Military Leave (calendar days)
CL — Conference Leave
JD — Jury Duty
SV — Vacation from Sick
MV — Management Vacation
CT — Comp Time

Department ______________________

______ Days ______ from ______________ to
Code

______________________ (inclusive dates).

______ Days pay in advance requested on

______________________ (last shift worked).
Date

Regular pay checks of: ______________

____

Explanation ______________________

____

STATION OFFICERS SIGNATURE NECESSARY FOR SICK LEAVE.

DISAPPROVED
APPROVED ______________________
Station Officer

DISAPPROVED
APPROVED ______________________
Commander

DISAPPROVED
APPROVED ______________________
Chief Officer

FORM AA-108 Rev. 4/82

Figure 5.3 Typical preprinted leave request form (Courtesy of Fort Lauderdale, Florida, Fire Department)

Format

The organization of data in each type of report follows fairly uniform formats. The purpose and audience of your report will determine if you will present your report in a memo or letter, or in a separate, titled report. If the report is lengthy, you will want to consider a cover letter, an abstract, table of contents, and appendixed exhibits. These elements will be discussed in Chapter 6. Chapter 1 listed visual devices which will make your message more readable. Devices such as topical headings, capital letters, underlining, variable spacing, numbers, and bullets can enhance even the briefest report. Make your report convey its message as clearly and concisely as possible.

Graphics

Besides planning your format (organization, headings, and other visuals) you should consider incorporating graphics to present your data. Study your message to determine if informal tables, formal tables, circle graphs, simple drawings, maps, charts, and other graphics will present the information more clearly than lengthy paragraphs of explanation. An incident report detailing a work-related accident might include a brief location diagram and an informal table of repair or replacement costs. A progress report may call for a formal table of materials, costs, or scheduling dates. A lab or test report will best display testing results in tables or performance curves. A feasibility report may include the results of surveys presented in circle graphs, bar charts, or tables. Further, tables comparing features of optional equipment may be appropriate. Be alert to graphic possibilities.

Language

As in all occupational writing, your language should be factual and objective. Avoid opinionated or judgmental language. Consider this biased paragraph from an accident report:

> A minor accident occurred at the Broward Community College Central Campus Library recently. Librarian Neil Springer carelessly overturned a cup of coffee on the microfilm machine which resulted in a massive short circuit when the machine was turned on. Luckily, there were no personal injuries.

The emphasis here is on blame rather than on the incident itself. An improved version would emphasize the facts:

> An explosion occurred on the second floor, east wing of the Broward Community College Central Campus Library on June 13, 198X, at 3:42 P.M.
>
> The cause of the explosion was an overturned cup of coffee spilled by Librarian Neil Springer onto an Acme 710 microfilm projector (Inventory #703-11201-8). The liquid penetrated the felt light seals and dripped onto the main power supply resulting in an explosive short circuit when the machine was turned on by student Mary Phillips.
>
> Although there were no personal injuries, damages to the machine and surrounding area total $480.00.

The revised version contains more essential facts and presents them in an unbiased manner.

INCIDENT REPORTS

Purpose

No matter where you work, the unexpected frequently occurs. Such digression from normal operating procedure generally requires an incident report to supervisors or others to prevent the incident from recurring. The incident report is a written investigation of accidents, machine breakdowns, delivery delays, cost overruns, production slowdowns, or personnel problems.

The incident report may be reviewed when the next budget is planned if your recommendations involve finances. The report may constitute the basis of a longer proposal to improve procedures. It may even be used as legal evidence in a follow-up investigation. A carefully detailed report becomes part of the written record of what goes on in your place of work.

Organization

The incident report adheres to fairly conventional organization. Its parts cover

- What happened (factual, not opinionated)
- What caused it (detailed and chronological)
- What were the results (injuries, losses, delays, costs)
- What can be done to prevent recurrence (recommendations)

In a lengthy incident report it is a good idea to include topical headings, such as

Incident		Accident Description
Cause	or	Analysis of Causes
Results		Corrective Action
Recommendations		Recommendations

Your reader will be able to cull the appropriate information quickly by glancing at the headings.

Incident description. In your introductory material, write a concrete statement detailing what happened. Include the exact date, time, and location. Personnel details, such as employees' names, titles, and departments, should be included. If personal injuries occurred, include the name(s) of victim(s), titles, and departments, or in the case of victims who are not employees, include home addresses, phone numbers, and places of employment. Describe the actual injury. If equipment is involved, identify it by including brand names, serial numbers, inventory numbers, or other pertinent descriptive detail.

Analysis of causes. In this section write a chronological review of what caused the incident. Include what was happening prior to the incident and each step which caused the incident.

Results. In this section, explain what happened due to the incident, such as the action which was taken immediately. If anyone was injured, describe the extent of the injury and how, when, and where the person was treated. Explain who was immediately involved. This may include paramedics, police, repair experts, or extra workers. In the case of equipment failure, explain how it was repaired or replaced, how late deliveries were speeded, or how high costs were curtailed. Detail what was done to settle a personnel problem or to satisfy a customer demand.

The results section may require an actual or estimated expense review. Include a precise breakdown of medical expenses, equipment replacement, repair costs, profit loss, or other applicable costs.

Recommendations. This section should include concrete suggestions to prevent the incident from recurring. Consider what should be done, who should do it, and when it should be done. Include as much detail as your position authorizes.

Figure 5.4 shows a printed accident report which illustrates the conventional organizational format.

Figure 5.5 shows the narrative portion of a fire incident report. Page one of the report, which is not included in the sample, is a printed check-list which serves as a record of location, method of alarm, fire origin, extent of damage, and other pertinent details.

Figure 5.6 shows a brief incident report illustrating logical organization and detail. The incident, collapsed shelving and broken merchandise,

OCCUPATIONAL INJURY AND ANALYSIS REPORT
ABG 70568 OPR-150 REV. 6/81

EASTERN

CAUTION — NCR PAPER
PLEASE PRINT - COMPLETE IN DETAIL

Refer to SP 52-202

REPORING DIVISION: ☐ MAINTENANCE ☐ SALES & SERVICE ☐ OTHER

EMPLOYEE DATA

EMPLOYEE NAME - LAST | FIRST | M.I. | EMPLOYEE NO. | DATE OF BIRTH | SEX | SOCIAL SECURITY NO. | EMPLOYMENT DATE | BASE STATION

STREET ADDRESS | CITY | STATE | ZIP CODE | HOME PHONE NO./AREA CODE | JOB CLASS | COST CENTER NAME | COST CENTER NO.

NO. DEPENDENTS UNDER 18 | MARITAL STATUS ☐ SINGLE ☐ MARRIED ☐ DIVORCED ☐ WIDOWED | CONSECUTIVE DAYS ON DUTY | SCHEDULED DAYS OFF | DATE OF INJURY | TIME OF INJURY

LENGTH OF TIME ON DUTY ☐ 1 HR. OR LESS ☐ 1-4 HRS. ☐ 4-8 HRS. ☐ OVER 8 HRS. | DAY OF WEEK | REPORTED TO SUPERVISOR DATE: ______ TIME: ______ | INJURY PAST TWELVE MONTHS ☐ 1st. ☐ 2nd. ☐ 3rd. ☐ 4th. ☐ RECURRENT ☐ MORE

NAME OF IMMEDIATE SUPERVISOR | TITLE | COMAIL ADDRESS | EXTENSION NO.

SUPERVISOR'S INVESTIGATION

1 - PART OF BODY AFFECTED
☐ HEAD/FACE ☐ LEG ☐ NECK
☐ HAND/FINGERS ☐ EYE ☐ ELBOW
☐ FEET/TOES ☐ WRIST ☐ CHEST
☐ SHOULDER ☐ KNEE ☐ OTHER (specify)
☐ BACK ☐ ANKLE

2 - TYPE OF INJURY
☐ BRUISE ☐ FRACTURE
☐ BURN ☐ CUT ABRASION
☐ SPRAIN ☐ MULTIPLE INJURIES
☐ RASH ☐ AMPUTATION
☐ HERNIA ☐ OTHER (specify)

3 - SPECIFIC TASK
☐ PUSHING ☐ SLIPPING ☐ STRUCK AGAINST OBJECT
☐ PULLING ☐ TRIPPED ☐ STRUCK BY OBJECT
☐ LIFTING ☐ FALL FROM ELEVATION ☐ OTHER (specify)
☐ KNEELING ☐ CAUGHT BETWEEN
☐ REACHING

4 - EQUIPMENT INVOLVED
VEHICLE NO. ______ ☐ PORTABLE STEPS
☐ BENT LOADER ☐ DOLLIE
☐ LOADER LIFTER ☐ CLEANING VAN
☐ FUEL TRUCK ☐ POWER CART
☐ FUEL PUMPER (Hydrant) ☐ JETWAY
☐ CONTAINER TYPE ☐ TRACTOR
______ ☐ COVERED CART
☐ V BED CART ☐ OTHER (specify)
☐ LAV TRUCK

5 - AIRCRAFT INVOLVED
AIRCRAFT NO. ______
☐ A300
☐ DC9
☐ B727
☐ L1011
☐ OTHER (specify)

6 - LOCATION OF ACCIDENT
☐ BAG ROOM ☐ ACFT REAR STAIRS
☐ BAG ROOM BELT ☐ ACFT FWD STAIRS
☐ AIR FREIGHT ☐ ACFT CABIN
☐ LOADING DOCK ☐ ACFT BELLY
☐ RAMP ☐ FWD BIN
☐ ACFT GALLEY ☐ AFT BIN
☐ CLAIM AREA ☐ BIN 3
☐ TICKET COUNTER ☐ OTHER (specify)
☐ INSIDE VEHICLE/VAN

7 - OBJECT INVOLVED
WEIGHT OF OBJECT ______ lbs. ☐ INVALID PSGR
☐ LIQUOR KITS
☐ BAGGAGE ☐ LIQUOR PIGS
☐ COMAT
☐ FREIGHT ☐ COST CUTTER
☐ MAIL WGT. ______
☐ HUMAN REMAINS AMT. ______
☐ TRASH CONTAINER ☐ OTHER (specify)
______ GAL SIZE

8 — NARRATIVE - WHAT WAS EMPLOYEE DOING WHEN INJURED? DESCRIBE EVENTS AND NATURE OF INJURY FULLY. IF EQUIPMENT MALFUNCTIONED, EXPLAIN.

9 — CORRECTIVE ACTION YOU HAVE TAKEN OR WILL TAKE TO PREVENT RECURRENCE OF ACCIDENTS OF THIS TYPE

M.D. OR R.N. REPORT

☐ RETURNED TO WORK ☐ FIRST AID TREATMENT ONLY ☐ RETURNED FOR ADDITIONAL TREATMENT ☐ RETURNED TO WORK WITH LIMITATIONS (see remarks)
☐ TREATED BY PHYSICIAN ☐ NOT RETURNED TO WORK (see remarks) DATE: ☐ ADMITTED TO HOSPITAL

REMARKS: | NAME OF ATTENDING PHYSICIAN OR NURSE ☐ M.D. ☐ R.N. | (AREA CODE) TELEPHONE NO. | DATE

SUPV. FOLLOW-UP

DID EMPLOYEE RETURN TO WORK ON NEXT SCHEDULED SHIFT? ☐ YES ☐ NO | WITH MEDICAL LIMITATIONS ☐ YES ☐ NO | WAS EMPLOYEE TRANSFERRED TO ANOTHER JOB? ☐ YES ☐ NO | IF YES, WAS IT: ☐ TEMPORARY ☐ PERMANENT | ☐ I HAVE PERSONALLY INVESTIGATED THIS ACCIDENT AND VERIFY IT OCCURRED AS DESCRIBED

☐ I AM UNABLE TO VERIFY THIS ACCIDENT AS DESCRIBED BY EMPLOYEE FOR THE FOLLOWING REASONS: | REASON(S)

REVIEW

PREPARED BY (SIGNATURE) | TITLE | EXTENSION NO. | DATE
EMPLOYEE'S SIGNATURE | POSITION | EXTENSION NO. | DATE
MANAGER'S SIGNATURE | TITLE | EXTENSION NO. | DATE

DISTRIBUTION: ORIGINAL (white) — Local File (5 years) DUPLICATE (yellow) — Ground Safety, MIAGS TRIPLICATE (pink) — Worker's Compensation, MIAWC

Figure 5.4 Typical preprinted accident report

REMOVE CARBON

Station 2 District k Company E-2 10-8 1407 10-10 1438 Hi-Rise No

Mileage 5 Code 3 Out of City No Damage H M L N # Floors 1

Condition on Arrival: Smoke H M L N Flames H M L N

APPLIANCES USED	EXTINGUISHING AGENTS USED		HOSE LINES USED
______ Deluge Gun	______ Wetwater	______ Gals.	______ Booster Line
______ Deck Turret	______ Foam	______ Gals.	1 1½″ Pre-Conn.
______ Snorkel Gun	______ AFFF	______ Gals.	______ 1½″Pre-Conn-SP
______ Tele-Squirt	X Water	300 Gals.	______ 2½″ Wyed Lines
______ Ladder Pipe	______ Dry Chemical	______ Gals.	______ 2½″ Hand Lines
			______ 2½″ Master Str.

ON-SITE EXT. USED	ON-SITE SYSTEMS USED	
		______ 3″ Master Str.
______ Pressurized Water	______ Standpipe	1 2½″ Supply
______ Soda Acid	______ Sprinkler	______ 3″ Supply
______ Foam	______ Dry Chemical	
______ Dry Chemical	______ Carbon Dioxide	LADDERS USED
______ Carbon Dioxide	______ Halogenated	______ Aerial
______ Halogenated		______ Ground up-30′
		______ Ground over 30′
		______ Squirt

Personnel: Lt. J. Yancey, M. Clarke, W. McGill, D. LeValley

Man Minutes: 124

Operations: While assisting the Fire Prevention Division on a standpipe test at 1625 Southeast Tenth Avenue, a passerby advised of smoke coming from a building at the listed address. I advised dispatch to start a full response, and E-2 responded to same. Upon arrival, I found the building with heavy fire involvement to the north half. E-2 laid a line in and then extinguished the fire with our pre-connect. After total extinguishment, I found evidence of two separate areas of origin. I-24 and 34 came on the scene and made an investigation. E-2 returned to #2 Station.

Cause: Suspicious; found two origins of fire.

Damage: Medium fire damage; probable no dollar damage because the south half of the building had been gutted by a previous fire.

Note: For additional information see Alarm #821158, 3/1/8x.

Form AA-271 Rev. 4/81

Figure 5.5 ***Florida fire incident report (Courtesy of Fort Lauderdale, Florida, Fire Department; narrative portion only)***

MEMORANDUM

TO: Jacquelyn E. Steinberg, Director

FROM: Minnie L. Cooper, Intake Counselor M.L.C.

SUBJECT: Collapsed Shelving in Civil Rights Section Supply Room

DATE: September 23, 198X

On Monday, September 22, 198X, at 3:00 P.M. the top shelf collapsed in the civil rights section supply room, which is located opposite the entrance to the civil rights office. The supplies on the shelf were destroyed; however, no physical injuries were sustained.

At approximatly 2:45 P.M. Dorothy McReed, secretary, asked Michael Burney, human relations specialist, to place an inoperable typewriter in the supply room. Mr. Burney placed the typewriter in the only available space which was on the top shelf. The shelf was stocked with fifty reams of mimeograph paper and could not support the additional weight of a typewriter.

As a result, the shelf collapsed and broke ten bottles of black artistic ink and five bottles of cement glue. In addition, the spillage from the ink and glue destroyed five reams of erasable bond typing paper, three boxes of legal file folders, and three boxes of index cards.

The civil rights section budget has sufficient funds to replace the shelf and to replenish the supplies. The cost, $83.50, includes

$10.00	for one shelf to be built by Broward County Maintenance Division (no charge for installation),
20.00	for erasable bond typing paper (5 rm @ $4.00 ea),
18.00	for legal file folders (3 box @ $6.00 ea),
7.50	for index cards (3 box @ $2.50 ea),
3.00	for cement glue (5 six-oz btl @ $0.60 ea), and
25.00	for black artistic ink (10 three-oz btl @ $2.50 ea).

Figure 5.6 Brief incident report (Courtesy student Minnie L. Cooper)

In order to prevent accidents such as this from recurring, I recommend a.) supplies be rearranged with the heavier items on the bottom shelf and lighter ones on upper shelves, or b.) equipment and duplicating paper be stored in a separate room. Should you endorse the last recommendation, the now defunct ESAA Component supply room is available for such use.

MLC:jv

Figure 5.6 Continued

requires a repair and replacement breakdown, which is presented in an easy-to-read continuation table.

Figure 5.7 shows an incident report with topical headings and numbered recommendations. In each the message is complete, objective, and concise.

PERIODIC AND PROGRESS REPORTS

Purpose

A periodic or progress report provides a record of the status of a project over a specific period of time. The report or reports are issued at regular intervals throughout the life of a project to allow management and workers to keep projects running smoothly. A periodic or progress report states what has been done, what is being done, and what remains to be done. The report usually reviews the expenditures of time, money, and materials; therefore, decisions can be made to adjust schedules, allocate budget, or schedule supplies and equipment. Often a company employs preprinted forms to report routine progress. Employees simply fill in the blanks at the completion of tasks. Figure 5.8 shows a fire department's weekly and monthly Apparatus Report form.

Similar periodic reports may be required daily, weekly, quarterly, semiannually, or annually. They insure uniformity and completeness of data.

Organization

All narrative reports in a series of progress reports should be uniform in organization and format. A progress report covers

- A review of the aims of the project highlighting accomplishments or problems.
- A summary or explanation of the work completed
- A summary or explanation of the work in progress
- A summary or explanation of future work
- An assessment of the progress

Headings for the sections, however brief, are usually used. Consider these topical headings:

MEMORANDUM

TO: James Little, Director Administration & Contract

FROM: Jill Fowler, Personnel Specialist J.F.

SUBJECT: Assault on Susan Watson, 5/19/8X

DATE: May 22, 198X

INCIDENT

On Wednesday, May 19, 198X, at 9:35 P.M. a female employee of the Benefits Department, Susan Watson, was assaulted in the Ace parking lot by a white male who fled with her purse.

CAUSE

Ms Watson was approaching her car after working overtime when the attack occurred. According to Ms Watson, the suspect approached her from behind as she was nearing her car. Using a blunt object, he struck Ms Watson on the head, and she fell to the ground. The man then ran with her purse in the direction of Northwest Thirty-sixth Street.

RESULTS

Since Ms Watson's car keys were stolen with her purse, she walked back to the lobby. However, she could not gain entrance to the building because her access card was also in her purse. Bleeding profusely from her head wound, Ms Watson pounded on the door to gain entrance; however, the guard, Ralph Murchison, had temporarily left his post. Mr. Murchison returned to find Ms Watson unconscious on the pavement. He immediately called an ambulance, and she was taken to Palmetto General Hospital where she was treated for scalp lacerations and multiple bruises. She was held for observation overnight and was released on May 20. Her insurance will cover all medical payments. Mr. Murchison reported the incident to the police, but because Ms Watson is unable to describe her assailant, no arrest is expected.

***Figure 5.7** Brief incident report with headings (Courtesy of Jennifer Fernandez, Ryder System, Inc.)*

RECOMMENDATIONS

Several problems have been spotlighted due to this incident. First, women leaving the building after dark are extremely vulnerable to such attacks. There is no protection afforded to them, and the parking area is not well lighted. The security gates are not functional, and unauthorized persons can easily gain entrance to the grounds. There is no way for an employee in distress to make his or her presence known at the lobby door if the guard is away from the security desk. Therefore,

1. Two security personnel should be stationed outside of the building. One should be situated at the main gate until the card access system is perfected. Unauthorized persons should be turned away at this point. The other guard should patrol the parking lot by buggy and escort employees to their cars when summoned to the main lobby by walkie-talkie.

2. Proposals should be obtained to install better lighting in the parking area.

3. A buzzer or bell should be installed outside the lobby door to summon the inside guard in emergency situations.

4. The main security desk should never be left unattended. Your department should institute a new policy and procedure in this area.

5. Training should be developed in two areas:

    a) CPR and basic emergency medical training for security guards and

    b) Self-defense training for interested employees developed with the assistance of the police department.

I would like to meet with you to discuss my recommendations as soon as possible.

JF:jmf

Figure 5.7 Continued

APPAR. NO. _____ RADIO NO. _____ MO. _____ YEAR _____ CO. _____ MAKE _____

WEEKLY APPARATUS REPORT

	1st Monday	2nd Monday	3rd Monday	4th Monday	5th Monday
Check pumps—Record vacuum test					
Check pumps—Record pressure test					
Were all drains flushed?					
Did connections or packing glands leak?					
Does relief valve or governor operate satisfactorily?					
Change Hurst Tool Fuel					
Check and flush foam pick-up					

	1st Tuesday	2nd Tuesday	3rd Tuesday	4th Tuesday	5th Tuesday
Resuscitator—Blood pressure of lowest cylinder					
Air Chisel—Record pressure of lowest oxygen cylinder					
Demand Regulator Mask—Record pressure					
Portable Spotlight—Running test					
Wench & Cord—Operating check					
First Aid Kits—Inventory check					
Hurst Tool & Compressor—Running test					
Tires—Record pressure					

Form AA-122 Rev. 8/77

Figure 5.8 Sample periodic report (Courtesy of Fort Lauderdale, Florida, Fire Department)

Introduction		Overall Goal
Work Completed		Work Completed
Work in Progress	or	Expenses Incurred
Work Remaining		Present Work
Appraisal		Future Work
		Conclusions

The major factor of a progress report is time. Other features, such as costs, materials, and manpower, may be incorporated into the appropriate sections or attached as support materials. If the reports are being prepared frequently during the course of a project, they are characterized by brevity—phrases rather than sentences and paragraphs. Figures 5.9, and 5.10 and 5.11 show progress reports in memo and letter formats. They range from brief phrase reporting to a more detailed sentence report which includes the cost analysis of an interior design project.

FEASIBILITY REPORTS

Purpose

You may be assigned to look into a new project—a new product, the development of a new program, a relocation, the purchase of new equipment—to determine the practicality of the project. The feasibility report presents the evidence of your investigation and analysis plus your conclusions and recommendations.

Typically, feasibility reports analyze data to answer specific questions:

- Will a given product, program, service, procedure, or policy work for a specific purpose?
- Is one option better than another option for a specific purpose?
- How can a problem be solved?
- Is an option practical in a given situation?

Figures 5.12, 5.13, and 5.14 illustrate feasibility reports which address themselves to these various questions.

Organization

The feasibility report usually includes:

MEMORANDUM

TO: Susan Niles, Law Department

FROM: Catherine Robertson, Executive Secretary C.R.

SUBJECT: Ace Company, Inc. Acquisition — Progress Report

DATE: May 12, 198X

INTRODUCTION

Here is a progress report on the Ace Company, Inc. acquisition.

WORK COMPLETED

- Stock purchase and Non-compete Agreements have been typed in final draft.
- Copies of drafts have been mailed to Seller by sky courier.
- Hotel and plane reservations have been confirmed by Zeta Travel Agency for arrival in Detroit on June 25 at 3:30 P.M.

WORK IN PROGRESS

- Limousine has been requested for transfer of all parties to Hilton Hotel; awaiting confirmation.

WORK REMAINING

- Meeting with the Seller and lawfirm to take place June 26.
- Drafts of Stock Purchase and Non-compete Agreements to be typed in final form for execution by all parties.
- Upon return, file on acquisition to be completed, and all documents to be filed as previously discussed.

Figure 5.9 Brief progress report (Courtesy of Cary Benavides, Ryder System, Inc.)

APPRAISAL

Although this acquisition was originally scheduled for a May 1 closing, all work is up to date, all parties have been notified of the delays, and all parties are scheduled for the June 25-26 closing. With the final execution of all documents, this acquisition will be concluded.

CR:js

Figure 5.9 Continued

Acme Advertising, Inc.
206 Lexington Avenue
New York, New York 10112
September 24, 198X

Mr. Bert Campbell
Widget, Inc.
2552 Washington Street
Youngstown, Ohio 37602

Dear Mr. Campbell:

As you requested the following is a progress report on the West Coast advertising campaign of your new product, Super Widget.

WORK COMPLETED
On September 23, 198X, thirty-second radio commercials began appearing in the following markets: Portland, Seattle, San Francisco, Los Angeles, and San Diego. As we agreed, the commercials aired on local sports shows on one station in each of those cities.

Immediate public reaction was so enthusiastic that the radio commercials were expanded to the San Jose, Reno, Spokane, Las Vegas, and Eugene markets.

WORK IN PROGRESS
On October 1, 198X, full page ads will appear in the Sunday morning sports sections of newspapers in the thirty largest West Coast markets.

Regent Stores has agreed to special promotions for Super Widget at all their retail outlets on October 5, 198X.

WORK REMAINING
Depending on the success of the first few weeks of sales of Super Widget on the West Coast, we will determine our best approach for national advertising. A target date for national advertising is November 15. A national television campaign is, of course, the best means to reach the consumer.

Figure 5.10 Sample progress report (Courtesy of Nancy R. Merrell, Ryder Truck Rental, Inc.)

Advertising for National Football League broadcasts is very expensive but will reach the majority of people who will be interested in the product.

APPRAISAL
So far our surveys reveal that the public is enthusiastic about the new product, and if the West Coast is any indication, Super Widget will be a financial success.

Very truly yours,

Nancy R. Merrell

Nancy R. Merrell
Account Executive

NRM:eq

Figure 5.10 Continued

JONES AND SMITH, INC.
DESIGN STUDIO
1 Plaza Park
Boca Raton, Florida 33344
May 12, 198X

Mr. and Mrs. James Little
3709 South Ocean Lane
Miami, Florida 32333

Dear Mr. and Mrs. Little:

As you requested here is a progress report on the interior design services which you contracted for on June 15, 198X.

WORK COMPLETED

Two 84 in. long, armless, three-cushion sofas with platform base and no skirt; poly and down combination; selected and ordered.

Twenty yd. custom print — gold mist brown/fluff red and gold on natural canvas — selected and ordered.

Two 24 in. x 24 in. x 20 in. lamp tables selected and ordered.

Two brown vinyl suede swivel lounge chairs selected and ordered.

WORK IN PROGRESS

The sofas are being constructed by Allied Upholstery Company, Fort Lauderdale, Florida. Estimated delivery date is July 29, 198X.

The lamp tables are being constructed by Joseph Walch, Fort Lauderdale, Florida. Estimated delivery is July 27, 198X.

I am collecting catalogs of lamps, coffee tables, and carpeting as we discussed.

WORK REMAINING

Two end tables, glass coffee table, and living room carpeting to be selected.

Figure 5.11 Sample progress report

Sofas, lamp tables, and chairs to be delivered.

Track lighting for ceiling to be installed.

APPRAISAL

All work is progressing on schedule. Following is a table of estimated and actual costs:

ESTIMATED AND ACTUAL COSTS FOR LITTLE FURNISHINGS
$6000.00 BUDGET

Item	Estimated Cost ($)	Actual Cost ($)	Studio Fee(30%) ($)	Florida Tax(5%) ($)	Total ($)
Sofas	2000	1850	555	120.25	2525.25
Material	1000	700	210	45.50	955.50
Tables	600	396	119	25.74	540.74
Chairs	700	490	147	31.85	668.85
Lamps	400	–	–	–	–
Coffee table	600	–	–	–	–
Carpeting	500	–	–	–	–
Electrician	100	–	–	–	–

We have several lamp and coffee table selections to show to you which are well beneath the $1000 budgeted for them. I feel confident that we can make the final selections within the original $6000.00 budget.

We hope you will enjoy your two week vacation in Maine. Please contact us when you return on July 20 to verify delivery dates and to make the final selections.

Yours truly,

Marcey Jones

Marcey Jones

MJ:cc

Figure 5.11 Continued

- Explanation of the problem
- Preset standards or criteria
- Description of the item(s) or subject(s) to be analyzed
- An examination of the scope of the analysis
- Presentation of the data
- Interpretation of the data
- Conclusions and recommendations

A brief feasibility report does not require headings; however, for a longer report consider these headings:

Background		Introduction
Standards		Problem
Options	or	Criteria
Method		Options
Data		Limitations
Conclusions		Recommendations

Background. This section includes all introductory material, such as the purpose of the report, a description or definition of the question, issue, problem, or item(s). You may discuss the scope or extent of the report.

Standards. Here you present a detailed explanation of the established criteria, aims, or goals of the question being investigated.

Options. If applicable, present each alternative according to your established criteria. Consider costs, capabilities, procedures, personnel involved, required training, or other appropriate features of each option.

Method. In this section explain how each option was analyzed to determine its practicality. This may include description of testing methods, survey instruments, research source material, qualifications of consultants, and discussion of limitations to your investigation.

Data. In this section present the test results, survey results, or research findings.

Recommendations. Finally, the feasibility report summarizes the investigation, drawing logical conclusions. Here you interpret the data and offer your recommendations.

Figure 5.12 shows a brief feasibility report in a memo format. It analyzes whether product X or product Y is better for the particular needs of a department in a large firm. Although it does not contain headings, the message covers the *background*, the *criteria* established for replacing

MEMORANDUM

TO: J. D. Big, Director, Accounts Department

FROM: Jane White, Secretary, Accounts Department J.W.

SUBJECT: Replacing Copy Machine

DATE: May 9, 198X

Inasmuch as we have been experiencing difficulties with our Atlas copy machine, I have investigated the feasibility of our renting a new copier. The Atlas costs $311.00 per month to rent. It does not make two-sided copies. In the past 30 days we have needed seven service calls costing a minimum of $30.00 a call. In April we were without copy capability for three days while a part was being located by the service representative.

Our criteria for a new machine include

- low cost (below $400 a month)
- two-side print capability
- various paper quality capability
- reduction capability
- immediate delivery

Only two copy machines are available within our price range. Table 1 compares the costs, capabilities, and limitations of Brand X and Brand Y:

Table 1 Comparative Features of Brand X and Brand Y Copy Machines

Brand	Rental Cost Per Month ($)	Capabilities	Limitations
X	316.00	• One step operation for two-sided print • Any paper • Reductions • Immediate delivery	• $35.00 base service fee • Reputation for frequent breakdowns
Y	389.00	• One step operation for two-sided print	• $75.00 base service • Reputation for slow

Figure 5.12 Sample feasibility report

- Any paper — delivery
- One-size reductions
- Immediate delivery

Both copiers meet our criteria; the Brand Y one-size reductions are suitable. In order to assess the disadvantages, I surveyed five departments which use the X or Y copiers to determine the number of service calls required in a one year period. Table 2 shows the departments and number of service calls required for each:

Table 2 Service Calls for Brand X and Y Copy Machines in One-Year Period

Department	Brand	# of calls (one year)
Personnel	X	11
Payroll	Y	2
Data Processing	X	25
Records	X	7
Purchasing	Y	0

Both Brand X and Brand Y appear to be more reliable than our Atlas copy machine. Although Brand Y charges more ($75.00) for a base service fee than does Brand X ($35.00), the survey data suggests that Brand Y is the more reliable machine.

Therefore, I recommend that we rent a Brand Y copy machine which will give us a two-side print capability which we do not presently have as well as meet all other criteria.

May I have your authorization by Friday to negotiate a Brand Y rental?

JW:eg

Figure 5.12 Continued

a copy machine, a review of the *optional equipment* capabilities, the *method* of weighing the alternatives along with the survey data, and a *recommendation* for action.

Figure 5.13 shows another feasibility report. This report analyzes the practicality of initiating a new academic program at a community college to meet the demands of local businesses and industries. Again the organization is basically the same. Although the headings are tailored for the particular report, the areas covered are the *introduction*, the *problem*, the *criteria*, the new *option*, the *method* of determining its feasibility, the available *data*, and *recommendations*.

Figure 5.14 shows a third feasibility report. This report addresses itself to solving a problem of abuses in a present bonus program. The organization remains much the same; the report presents the *background* and *problem*, the *standards* for an equitable bonus plan, the details of an *optional plan* along with its possible *limitations*, and the *recommendations* for adopting the new bonus program.

LABORATORY AND TEST REPORTS

Purpose

Students and employees in the fields of chemistry, data processing, fire science, electronics, nursing, and other allied health areas must frequently write laboratory or test reports. The reports present the results of research or testing.

Organization

Typically the report includes

- Statement of purpose
- Review of method or procedure of testing
- Results
- Conclusions and recommendations

A test report is usually less formal than a true laboratory report. The test report may be transmitted in a memorandum or business letter. Figure 5.15 shows a test report from an automotive laboratory which tested the durability of fiberglass impellers for use in automobile water pumps. It is clear that the *purpose* of the tests was to determine the endurance time of the impellers. The *procedure* used to conduct the test is described only to the extent that it would interest the reader. The

MEMORANDUM

TO: Mary Ellen Grasso, English Department Chair
All Instructors

FROM: Eric Delaney, English Instructor E.D.

SUBJECT: Technical Report Writing Certificate Program

DATE: February 15, 198X

INTRODUCTION

The purpose of this report is to examine the feasibility of initiating a Technical Report Writing Certificate Program, a multi-discipline curriculum to train students to communicate technical and/or occupational factual data in a grammatical, concise, objective manner enhanced by professional presentation and graphic illustration in tandem with the practical demands of business and industry.

BACKGROUND

Representatives from several local businesses and industries (Bendix, Motorola, American Express) have expressed concern that the graduates of the college who apply for technical writing positions lack the skills required by technical writer job descriptions. Although the College offers courses in occupational writing and technical and professional writing, it does not offer an organized program covering technical writing, graphics, and business communications.

An instructor exchange program has been instituted wherein one instructor from the college teaches technical writing on the premises of an industry while an industrial technical writer teaches part-time at the college. This exchange is piecemeal at best. A comprehensive program of training students for industrial writing positions is still needed.

Figure 5.13 Sample feasibility report

CRITERIA

Business and industry requires writers who have the following skills:

1. a command of grammer, punctuation, and usage;
2. the ability to organize information logically and to express data clearly and concisely;
3. a grasp of business letter, memo, report, proposal, and technical manual writing requirements;
4. a capability in graphic preparation;
5. the ability to express oneself orally, and
6. the fundamentals of typing.

SUGGESTED PROGRAM

To meet the requirements the following 21-hour course outline for a Technical Report Writing Certificate Program should be considered:

Course	No. of credits
Grammar	3
Composition I	3
Occupational Writing	3
Composition II. Pre-professional Writing	3
Basic Typing, Part I*	1
Basic Typing, Part II*	1
Basic Typing, Part III*	1
Introduction to Speech Communications	3
Graphic Communications	3
Total Credit Hours	21

*An equivalency test may be substituted for 3 credits

In addition to regular daytime scheduling, courses should be offered from 5:00 to 10:00 p.m. in order to accommodate students who are already employed in business and industry.

METHOD

To validate the feasibility of the program we need to

Figure 5.13 Continued

1. survey similar programs at other Florida community colleges,
2. establish an advisory board,
3. meet with instructors in speech and engineering technology concerning the development of appropriate courses,
4. submit the program to the Academic Affairs Committee for approval, and
5. advertise the program widely to business, industry, and the public.

SURVEY

A survey of ten Florida community colleges reveals that six institutions offer similar certificate programs. They range from 18 credit to 24 credit requirements. Two do not require three credits in grammar, two require three credits in standard composition II. Enrollment in one program is limited to men and women in business and industry whose companies pay the tuition.

ADVISORY BOARD

I recommend the board be comprised of eleven persons: the chair of the English Department, the Chair of the Engineering Technology department, one instructor in speech, two instructors in English, and six representatives from the business and industrial community.

INSTRUCTORS

Although the courses in grammar, composition, occupational, and preprofessional writing, and typing are already offered, new courses in Introduction to Speech Communications and Graphic Communications will need to be developed. Discussions with George Cavanagh, Chair of the Speech Department; and Dr. S. L. Oppenheimer, Director of Engineering Technology, indicate that such courses are feasible.

Figure 5.13 Continued

RECOMMENDATIONS

1. The certificate program will meet the needs of students seeking entry-level employment in technical writing and students working in business and industry who wish to upgrade their professional skills.

2. Course outlines for Introduction to Speech Communications and Graphic Communications should be submitted by June 15, 198X.

3. The advisory board should meet twice this semester to discuss the program and the curriculum. Suggested dates are March 15 and May 15.

4. The program should be advertised in the catalog, in local media, and by 5000 brochures to local business and industry.

5. The program should be implemented by Term I, 198X–8X.

ED:jsv

Figure 5.13 Continued

NORTHEAST REGION MCS

BONUS PROGRAM

Feasibility Report

I. INTRODUCTION

The Northeast Region MCS Partners are considering the feasibility of discontinuing annual bonuses and of paying bonuses to professional staff on an individual basis throughout the year immediately following bonus-worthy events or conditions. In this way the bonus would be kept separate from salary considerations, and the reward would more closely relate to the event or condition.

II. BACKGROUND

For many years the Northeast Region MCS has had a bonus program for professional staff. Each staff member was eligible for an annual bonus payable September 30. Whether or not he was paid a bonus and what the amount would be were determined at the time of the June performance evaluation. When paid, the bonuses ranged up to 15% of annual salary.

The bonus was meant to recognize and reward unusual contribution or difficult conditions during the previous year. It also reflected the economic condition of the practice. In good times, total bonus payments were higher than they were in bad times. In one recent year, when operations were showing a loss, there were no bonuses.

The program was abused in two ways. To some extent partners and staff began to consider bonuses as regular, recurring payments with only the amount subject to annual determination, and they sometimes were used as a substitute for salary increases that were not as permanent as a salary increase. Also, during wage controls, some payments that would have been salary increases in other times were awarded as bonuses.

Figure 5.14 Sample feasibility report (Courtesy of Charles E. Smith, Jr.)

III. STANDARDS

An equitable bonus plan must

1. motivate staff toward desirable activities,
2. reward unusual contribution,
3. compensate for difficult conditions, and
4. make salary adjustments more representative of a staff member's overall performance.

IV. OPTIONAL BONUS PLAN

Under the new program, bonuses will be awarded for the same reasons as before—unusual, meritorious contribution and undue hardship. Some examples follow:

A. Meritorious Contribution

- For any professional, a contact or other development work leading to securing a new audit client.
- For a manager, a self-conceived practice development program of self-initiated contact leading to a consulting engagement. (Bonuses should not be awarded to managers for an excellent sales record per se.)
- For any professional, the conception and eventual use of a unique approach to an engagement-related work task.
- For any professional, the development of a technique (engagement-related or not) that has wide application in our practice.
- For a staff consultant, completion of a clearly defined engagement work task in significantly less time than had been budgeted by a partner or manager.
- For a staff consultant, a contact or other development work leading to securing a consulting engagement.

Figure 5.14 Continued

B. Undue Hardship

- An extended period of travel away from home, defined as spending more than 75 percent of weekday nights away from home over a six-month period or spending one-half of the weekends in a four-month period away from home.

- A close working relationship for a month or more with intransigent, unreasonable, abusive, or otherwise difficult client personnel.

- Uncomfortable physical working conditions for a month or more, such as at a remote, isolated community or in a noisy, dirty, hot, or cold facility.

- An extended period of weekend or evening work where total hours worked are more than twice normal hours for a month or more.

V. PROBLEMS

This bonus plan option may present some problems. Some members of the staff, both managers and consultants, are apt to direct their efforts towards earning bonuses, and thereby pay less attention to their regular professional work. Furthermore, they may embarrass the Firm in doing so.

For example, a consultant may devote too much time on an engagement looking for a unique approach rather than following the work plan as laid out by the manager or partner. It is also possible that a manager or staff consultant may actively seek new audit clients at the expense of MCS engagements, or his activity in developing an audit client may conflict with plans of the general practice or infringe on professional ethics. There is also a remote possibility that a member of the professional staff might try to encourage rather than allay intransigence in a client.

Figure 5.14 Continued

VI. RECOMMENDATIONS

1. The optional bonus plan should be adopted for a period of two years because it provides for more equity than the previous bonus plan.

2. The Northeast Region MSC Partners should develop preliminary guidelines listing events and conditions and the amounts of bonus merited by each.

3. To provide initial equity any partner may propose a bonus for any staff member who may warrant a bonus within the next six months. The bonus proposal will be discussed at each scheduled partners' meeting and approved or disapproved by the Regional Director. In this way the guidelines may be further refined, and all of the partners will develop a consistent point of view.

4. Subsequently, bonuses will be approved or disapproved directly by the Regional Director without joint discussion, but the justification for the amounts of all bonuses paid in the preceding period will be presented for informational purposes at each partners' meeting.

CHS:eg

Figure 5.14 Continued

UNIVERSAL TESTING LABORATORIES
1111 North University Drive
Plantation, Florida 33313

October 6, 198X

Mr. Thomas Jones
5445 Northwest Third Court
Plantation, Florida 33313

Dear Mr. Jones:

Thank you for the submission of the fiberglass impellers to us for testing as an alternative to cast iron impellers for automobile water pumps. To determine the endurance of your impellers, we destruct tested the impellers under simulated load conditions at various revolutions per minute (RPM). The table indicates the endurance times of the test impellers in comparison to the standards set by the American Automobile Association:

FIBERGLASS IMPELLER ENDURANCE TEST AT VARIABLE RPM

Test Number	RPM	Endurance Time (Minutes)	American Automobile Standards (Minutes)
1	4000	248	300
2	5000	242	275
3	6000	235	250
4	7000	225	225
5	8000	206	200
6	9000	165	175
7	10000	75	150

This same information is illustrated in a curve:

Figure 5.15 Sample test report (Courtesy Cindy E. Grotsky)

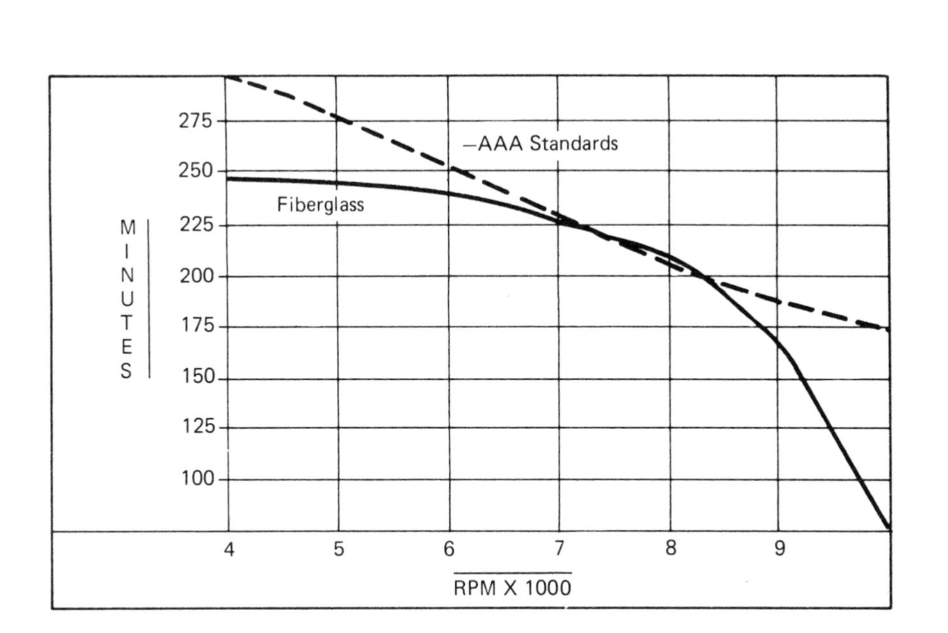

It is apparent from the data that the fiberglass impeller does not meet the minimum standards of endurance set by the American Automobile Association, except at 7000 and 8000 RPM. It would not be practical to produce the fiberglass impeller at this time. We recommend that you reinforce the impeller with steel webbing for further testing.

We look forward to assisting you in your future refinements.

Very truly yours,

Cindy E. Grotsky

Cindy E. Grotsky
President

CEG:eg

Figure 5.15 Continued

results of the test are presented in graphics for clear comprehension, and the *conclusions and recommendations* are stated with reference to predetermined standards. Use topical headings if they will make your report clearer.

Laboratory reports generally cover the same four considerations, but are more formally structured. Figure 5.16 shows a report on the preparation of aspirin. It includes flow charts of the main chemical reactions, potential side reaction, and the separation scheme. The procedure section of the report details the method used to prepare the aspirin with attention to the equipment used. The results and conclusion are covered under separate headings.

FIELD REPORTS

Purpose

A field report presents an analysis of a location, site, or situation to record and determine appropriate action. Realtors prepare field reports on undeveloped, commercial, and industrial properties to determine the value and prospects of such properties. Service people from every field inspect property to determine costs and plans for building on, improving, or repairing the property. Fire fighters, health inspectors, and others report their work in the field. Insurance adjustors inspect sites to establish claims for damages.

Organization

Often preprinted forms with space for narrative reporting are used. Many companies devise organization formats in outline form to be followed by their reporting personnel. These regulating outlines list specific topical headings along with brief instructions about the data to be included under each.

Despite the diversity of headings required for any one specific field report, all such reports include

- Essential background data
- Account of the field inspection
- Analysis of findings
- Conclusions and recommendations

Figure 5.17 shows a fire department company run report. Page one of the report records the essential background data while page two contains a narrative report about what happened in the field.

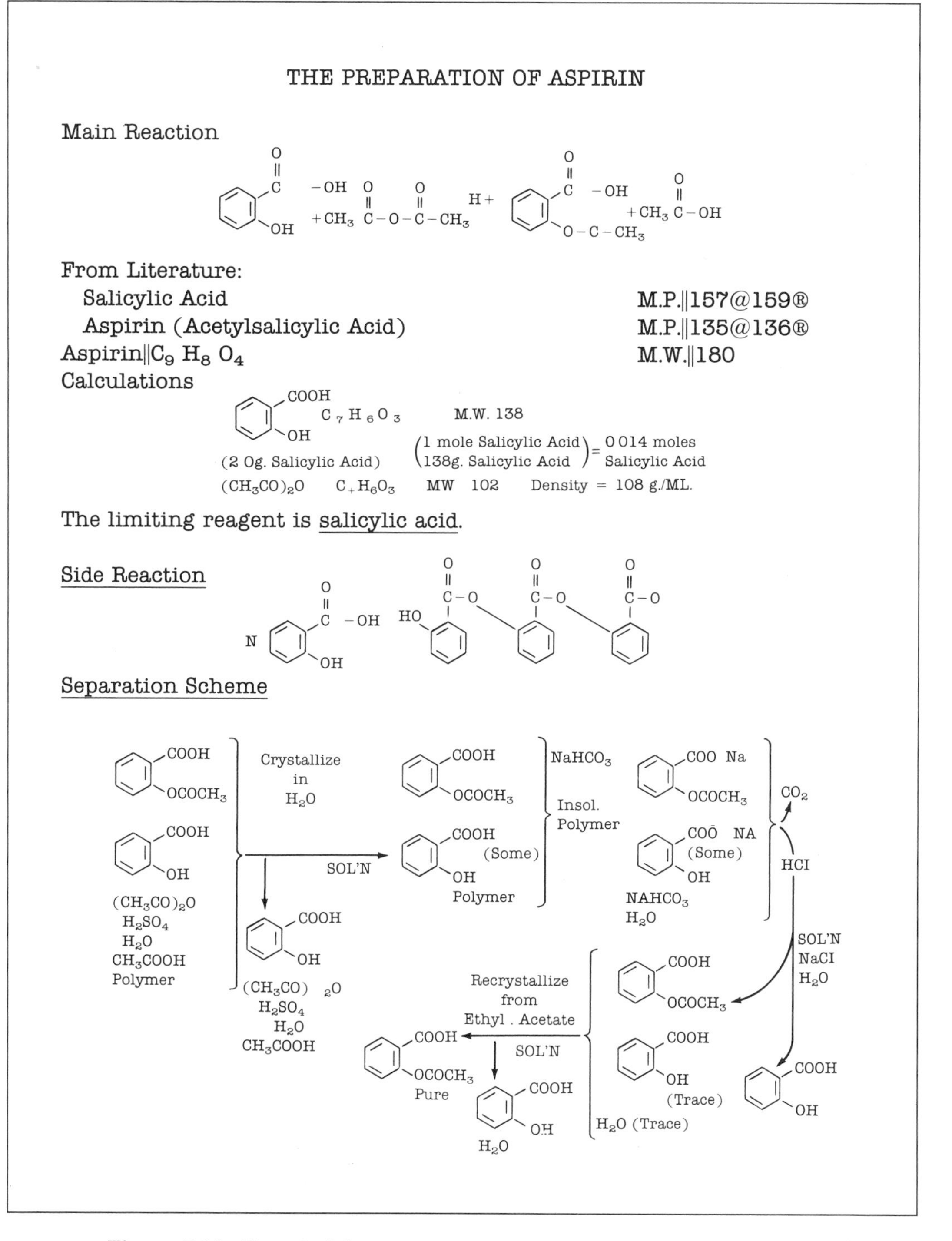

THE PREPARATION OF ASPIRIN

Main Reaction

$$\text{(salicylic acid)} + CH_3\overset{O}{\overset{\|}{C}}-O-\overset{O}{\overset{\|}{C}}-CH_3 \xrightarrow{H^+} \text{(aspirin)} + CH_3\overset{O}{\overset{\|}{C}}-OH$$

From Literature:

Salicylic Acid M.P.||157@159®

Aspirin (Acetylsalicylic Acid) M.P.||135@136®

Aspirin||$C_9\ H_8\ O_4$ M.W.||180

Calculations

$C_7H_6O_3$ M.W. 138

(2 0g. Salicylic Acid) $\left(\frac{1 \text{ mole Salicylic Acid}}{138g. \text{ Salicylic Acid}}\right)$ = 0 014 moles Salicylic Acid

$(CH_3CO)_2O$ $C_4H_6O_3$ MW 102 Density = 108 g./ML.

The limiting reagent is salicylic acid.

Side Reaction

Separation Scheme

Figure 5.16 ***Sample laboratory report (From Laboratory Manual for General Chemistry by Kenneth D. Whitten and Kenneth D. Gailey. Copyright 1981 by Saunders College Publishing. Reprinted by permission of Saunders College CBS College Publishing.)***

Procedure:

Salicylic acid: Sample + Paper 3.12g. $\frac{2.00g}{138g./mole}$ = 0.014 moles salicylic acid

Paper 1.12g.

Sample 2.00g.

The salicylic acid was placed in a 125 ml Erlenmeyer flask. Acetic anhydride (5 ml) was added along with 5 drops of conc. H_2SO_4. The flask was swirled until the salicylic acid was dissolved. The solution was heated on the steambath for 10 minutes. The flask was allowed to cool to room temperature, and some crystals appeared. Water (50 ml) was added, and the mixture was cooled in an ice bath. The crystals were collected by suction filtration, rinsed three times with cold H_2O, and dried.

Crude yield: Product + paper 3.07g.

paper 1.15g

1.92g. aspirin

Theoretical yield = (0.014 moles) (180g. Aspirin/mole) = 2.52g.

Actual yield = 1.92g.

Percentage yield = 76%

The crude product gave a faint color with $FeCl_3$. Phenol and salicylic acid gave strong positive tests.

The crude product was placed in a 150 ml beaker, and 25 ml of saturated NaHCO was added. When the reaction had ceased, the solution was filtered by suction. The beaker and funnel were washed with CA. Two ml of H_2O dilute HCI was prepared by mixing 3.5 ml conc. HCI, and 10 ml H_2O in a 150 ml beaker. The filtrate was poured into the dilute acid, and a precipitate formed immediately. The mixture was cooled in an ice bath. The solid was collected by suction filtration, washed three times with cold H_2O, and placed on a watch glass to dry overnight.

Yield: Paper and product: 2.78g.

paper: 1.08g.

product: 1.70g.

Theoretical yield = 2.52g.

Actual yield = 1.70g.

Percentage yield = 67% mp. 133–135°C.

The solid did not give a positive $FeCI_3$ test. The final product (CA. 0.75 g) was dissolved in a minimum amount of hot ethyl acetate. Crystals appeared. The crystals were collected by suction and dried. MP. 135–136°C.

Figure 5.16 Continued

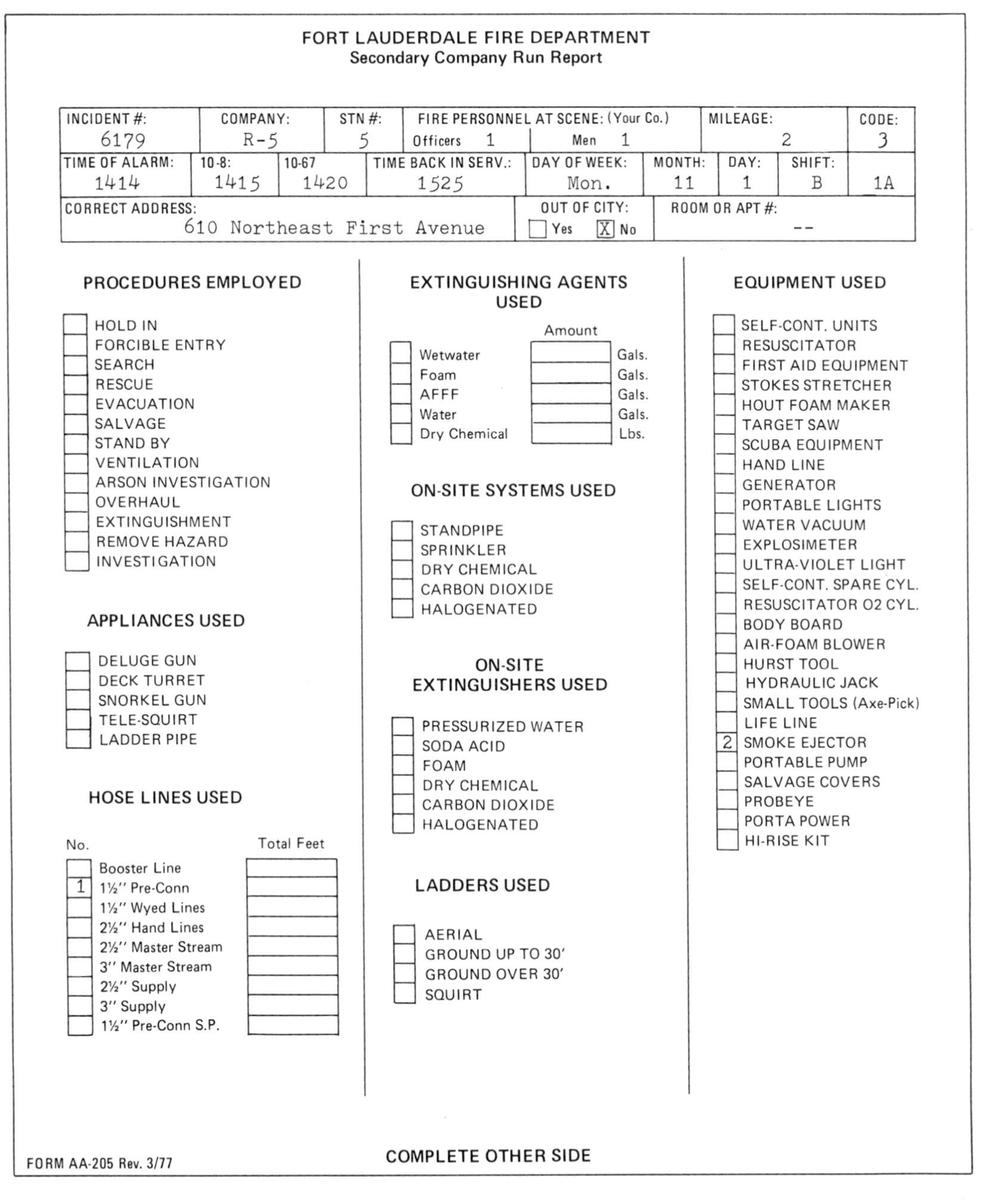

FORT LAUDERDALE FIRE DEPARTMENT
Secondary Company Run Report

INCIDENT #:	COMPANY:	STN #:	FIRE PERSONNEL AT SCENE: (Your Co.)		MILEAGE:	CODE:
6179	R-5	5	Officers 1	Men 1	2	3

TIME OF ALARM:	10-8:	10-67	TIME BACK IN SERV.:	DAY OF WEEK:	MONTH:	DAY:	SHIFT:	
1414	1415	1420	1525	Mon.	11	1	B	1A

CORRECT ADDRESS:	OUT OF CITY:	ROOM OR APT #:
610 Northeast First Avenue	[ ] Yes [X] No	--

PROCEDURES EMPLOYED

- [ ] HOLD IN
- [ ] FORCIBLE ENTRY
- [ ] SEARCH
- [ ] RESCUE
- [ ] EVACUATION
- [ ] SALVAGE
- [ ] STAND BY
- [ ] VENTILATION
- [ ] ARSON INVESTIGATION
- [ ] OVERHAUL
- [ ] EXTINGUISHMENT
- [ ] REMOVE HAZARD
- [ ] INVESTIGATION

APPLIANCES USED

- [ ] DELUGE GUN
- [ ] DECK TURRET
- [ ] SNORKEL GUN
- [ ] TELE-SQUIRT
- [ ] LADDER PIPE

HOSE LINES USED

No.		Total Feet
	Booster Line	
1	1½" Pre-Conn	
	1½" Wyed Lines	
	2½" Hand Lines	
	2½" Master Stream	
	3" Master Stream	
	2½" Supply	
	3" Supply	
	1½" Pre-Conn S.P.	

EXTINGUISHING AGENTS USED

		Amount	
[ ]	Wetwater		Gals.
[ ]	Foam		Gals.
[ ]	AFFF		Gals.
[ ]	Water		Gals.
[ ]	Dry Chemical		Lbs.

ON-SITE SYSTEMS USED

- [ ] STANDPIPE
- [ ] SPRINKLER
- [ ] DRY CHEMICAL
- [ ] CARBON DIOXIDE
- [ ] HALOGENATED

ON-SITE EXTINGUISHERS USED

- [ ] PRESSURIZED WATER
- [ ] SODA ACID
- [ ] FOAM
- [ ] DRY CHEMICAL
- [ ] CARBON DIOXIDE
- [ ] HALOGENATED

LADDERS USED

- [ ] AERIAL
- [ ] GROUND UP TO 30'
- [ ] GROUND OVER 30'
- [ ] SQUIRT

EQUIPMENT USED

- [ ] SELF-CONT. UNITS
- [ ] RESUSCITATOR
- [ ] FIRST AID EQUIPMENT
- [ ] STOKES STRETCHER
- [ ] HOUT FOAM MAKER
- [ ] TARGET SAW
- [ ] SCUBA EQUIPMENT
- [ ] HAND LINE
- [ ] GENERATOR
- [ ] PORTABLE LIGHTS
- [ ] WATER VACUUM
- [ ] EXPLOSIMETER
- [ ] ULTRA-VIOLET LIGHT
- [ ] SELF-CONT. SPARE CYL.
- [ ] RESUSCITATOR O2 CYL.
- [ ] BODY BOARD
- [ ] AIR-FOAM BLOWER
- [ ] HURST TOOL
- [ ] HYDRAULIC JACK
- [ ] SMALL TOOLS (Axe-Pick)
- [ ] LIFE LINE
- [2] SMOKE EJECTOR
- [ ] PORTABLE PUMP
- [ ] SALVAGE COVERS
- [ ] PROBEYE
- [ ] PORTA POWER
- [ ] HI-RISE KIT

FORM AA-205 Rev. 3/77

COMPLETE OTHER SIDE

Figure 5.17 Sample field report (Courtesy of Fort Lauderdale, Florida, Fire Department)

OPERATIONS AND COMPANY AND STORY OF FIRE

Give a complete account of the work of the company. The sequence of operations and where the company worked (floor of building, for example) should be clearly recorded. To the best of your recollection show where each man worked and what he did. Describe where fire apparently started and from what cause; what material burned; progress of fire upon arrival of company; and, any dangerous condition of the building. List all company personnel responding. List all injury and/or fatality information.

Condition of doors or windows (locked or open, for example); any unusual presence of flammable or combustibles; or, obstacles placed to hamper the fire department deliberately should be noted. This information should so far as possible be coupled with the name of the man who encountered the condition so that he can be witness to it if necessary in connection with further investigation of the fire.

Also, utilizing Form 901-J, make sketch to show the relative positions of pumper, hydrant, ladder or other apparatus with respect to the building afire. Show hose lines manned by this company and their approximate lengths.

R-5 Crew: Lts. Sicliri and Ferranti
Man Min: 140
Operation:

This unit was dispatched to a reported structure fire at topic address. Upon arrival on the scene District Commander One told me to bring a second line to the fire floor. We pulled an inch and one-half line from engine one S and advanced it to the fire floor. This line was not needed at this time. Driver Ferranti brought smoke ejectors and a pike pole to the apartment involved. DC-1 told me to check for fire extension. I went into the attic and crawled the length of the building checking for fire. Finding no fire in the area, I reported back to the Commander. Commander Zettek then had my unit stand by in reserve. When the area was safe, we cleared the area and returned to quarters.

By Howard J. Siliri Lt.
Officer in Charge

Checked By ____________

(Use Additional Sheet If Needed)

Figure 5.17 Continued

MEMORANDUM

TO: Ken Vordermeier, Vice-president

FROM: Jane Ellyson, Realtor J.E.

SUBJECT: Unimproved Land Inspection

DATE: September 22, 198X

On Friday, September 20, 198X, I inspected 3.4 + acres of unimproved land fronting on Powerline Road and Northwest Twentieth Street, Fort Lauderdale, Broward County, Florida, to determine its potential and to obtain the seller's listing.

DESCRIPTION

Legal - Hillmont Middle River Vista, Replat of a portion of Plat Book 59, Page 188, Parcel B less pt., Desc. in CRS 3197/876, 3195/834, 433/711, and 4638/335, less approximately 1.049 acres conveyed to McDonald's Corporation and less approximately 0.358 acres to be conveyed as Northwest Twentieth Street.

Style code: 97
Area code: 45
Tax number: 9228-13-002

Price - Listing price is $600,000.00. The property is free and clear, and the seller desires cash. The price per acre is $174,240.00 or $4.00 per square foot. Sale lease back is not available.

INSPECTION

This vacant property is 3.4 + acres, 150,000 + square feet, with 450 to 500 feet of frontage and irregular depth. The property is waterfront with the property line following river bank.

It is zoned for commercial land use and platted. No rezoning is required. A survey is available in the listing office. No improvements have been made.

Figure 5.18 Sample field report (Courtesy Vordermeier Company, Realtors; adaptation)

Minor clearing and/or grubbing, and fill will be required. The need for easements and restrictions are unknown at this time.

The property is accessible to Port Everglades waterport, Fort Lauderdale/ Hollywood airport, and the main highways: I-95, Florida Turnpike, and Oakland Park Boulevard. No railroad access is available. The property has access to electricity, city water, and sewers.

ANALYSIS

The surrounding area is developing rapidly. New Lake Lauderdale Park adjoins the property at the north end, and a fast-food restaurant is being built immediately south of the subject property.

The land offers an excellent prospect for a retail or service business.

RECOMMENDATIONS

1. Obtain the listing.

2. List on Realtron Computer service

Style code:	97 Commercial
General Search codes:	V2 Electricity
	V3 Public Water
	V4 Sewers or services
	V5 Paved streets or sidewalks
	V8 Major Road Frontage

3. Verify financing at time of contract.

Figure 5.18 Continued

MEMORANDUM

TO: Paul Austin, Superintendent, Ace Business Services, Inc.

FROM: Gloria Dunn, Field Claim Adjuster G.D.

SUBJECT: Fancy Fast Food Corporation - #43101-08459
Claimant Adele Clarke - Claim #172579

DATE: March 26, 198X

SUGGESTED RESERVE

Bodily injury - Adele Clarke - $10,000

FACTS

On 3/12/8X the claimant was a customer at the Fancy Fast Food premises located at 1688 South Ocean Boulevard, Miami, Florida. The claimant was apparently bouncing on a chair at station 7 nearest the kitchen when the left, rear leg of the chair broke, collapsing the chair and causing the claimant to strike her chin on the edge of the table in the ensuing fall.

LEGAL VIOLATIONS

There are no known legal or code violations.

DIAGRAM

Enclosed is a diagram of the restaurant interior and station 7 showing the location of the table and chair at the time of the incident.

PHOTOGRAPHS

Enclosed is a photograph of the table and broken chair.

INSURED

The insured is Fancy Fast Food's licensee, 1688 South Ocean Boulevard, Miami, Florida. The manager's name is Raymond Krick. We have met with

Figure 5.19 Sample field report

Mr. Krick, and he has indicated that this is the first incident involving a collapsed chair.

The chairs and other furniture were purchased and installed in February, 198X, by Restaurant Suppliers, Inc. of Miami, Florida. The rest of the tables, banquettes, and chairs appear to be in good condition, and no defects are noted. His attached statement is self-explanatory.

BODILY INJURY - Adele Clarke

The claimant is ten (10) years old and lives with her parents, Mary and Jerry Clarke, at 7210 West Seventh Street, Plantation, Florida.

Injuries

The claimant received two damaged teeth that were pushed completely up into her gums. The claimant was transported to Miami General Hospital emergency room where she was treated and then referred to an oral surgeon. The claimant will be having oral surgery including a bone transplant to correct the damage. The claimant's oral surgeon is Dr. M.C. Popper located at 1400 South Belvedere Street, Fort Lauderdale, Florida. Since the claimant is represented by an attorney, we were not able to obtain a statement or medical authorization from the claimant's parents.

Damages

The claimant's damages are unknown at this time; however, we expect the medicals and specials to be several thousand dollars.

CLAIMANT ATTORNEY

The claimant is represented by Attorney Julie King located at 609 North Andrews Avenue, Fort Lauderdale, Florida. We have spoken to Ms King and she is very cooperative and has indicated to us that the claimant seeks medical coverage because the Clarkes do not have any medical or dental insurance.

Although Attorney King did not make a definite statement about liability, she insinuated that a liability claim may be forthcoming.

Figure 5.19 Continued

WITNESSES

Enclosed is the statement of waitress Paula Kline, an employee of Fancy Fast Food, and it is self-explanatory.

The claimant's mother, Mary, was seated in the booth when the accident occurred, but since she is represented, we have not obtained her statement. There were no other witnesses to the occurrence.

RECOMMENDATION

At this time there appears to be a valid liability claim on behalf of Fancy Fast Food Corporation as well as claim to medical bills.

Once we have been able to obtain the medicals and specials from the claimant's attorney, I recommend we pay the medicals and attempt to obtain a Parent/Guardian Release for the amount of the medicals. Should a liability claim be pressed, I recommend we attempt compromise.

UNFINISHED ITEMS

1. Obtain medical and special costs
2. Determine if liability claim is forthcoming
3. Next report 4/30/8X

ENCLOSURES

1. F2-205 Form
2. Diagram
3. Photograph
4. Transcribed statement of Raymond Krick
5. Transcribed statement of Paula Kline

GD:jv

Figure 5.19 Continued

Figure 5.18 shows a realtor's field inspection report on a parcel of unimproved land. The data includes pertinent information required for a listing.

Figure 5.19 is a lengthier field report filed by an insurance adjustor regarding a claimant seeking medical expenses for an injury sustained while on the premises of an insured fast-food restaurant. The enclosures to the report are not reproduced in the sample.

EXERCISES

1. Use the following information to write an incident report in memorandum format. You are the front desk reservations employee reporting to the general manager. Use the appropriate format, organization, visuals, and graphics to make your report clear and readable.

Incident: Water pipe in ceiling burst; water damage to office equipment, carpet, and draperies.

When: 12:40 P.M., Friday, April 18, 198X

Where: General office, Holiday Hotel

Cause: Unknown

Damage: Smith Corona electric typewriter (serial No. 809-71-2654) soaked; requires cleaning and oiling.
Acme calculator soaked and short-circuited; requires replacement.
Carpet (20′ × 16′) needs replacement.
Draperies (6′ × 16′) need replacement.
No personal injuries.

Costs: Typewriter reconditioning—$35.00
Calculator replacement—$80.00
Carpet replacement—$512.00
Drapery replacement—$300.00
Plumber—$500.00

Further results: Estimated one-week delay in preparation of promotion mailout project.

Recommendations: Hire temporary typist for five days; estimated cost—$250.00

2. Use the following information to write a progress report in letter format. Arrange the data in sequential order. Use headings. Decide whether to present information in phrases or sentences.

Position: You are an Ace Company training specialist reporting to a writing consultant whose services have been contracted to conduct on-premises occupational writing instruction at your company.

Project: Writing Seminar to upgrade employee skills—March 5, 6, and 7, 198X.

Data: (follow directions from above):

1. Consultant evaluation forms to be devised.
2. Negotiating luncheon menus with James Watkins, cafeteria manager; all three dates.
3. Twenty registrants from 15 departments confirmed.
4. Writing samples from each registrant being solicited to forward to consultant for evaluation.
5. Training Room C scheduled 9:00–5:00 P.M., March 5, 6, 7.
6. Certificates of Completion to be typed.
7. Seminar announcements sent to 54 departments, February 1.
8. Xeroxing and binding of 20 sets of instructional materials in progress.
9. Overhead projector to be ordered for March 6, 9:00–5:00.
10. Wayne Salsbury to be notified to prepare introduction of Consultant for March 5.
11. Memos to supervisors explaining cost center billing sent on February 15, 198X.
12. Tables to be arranged in semicircle for registrants on March 5, 6, 7.

3. Use the following information to write a feasibility report in memorandum format. You are the Grounds Committee Chairperson reporting to the Golden Lakes Condominium Association. Rearrange the data into logical organization. Present the data in appropriate graphics. Based upon the information, draw logical recommendations in your conclusion.

Problem: Remodeling of Golden Lakes Condominium recreation building has resulted in grass damage in common areas.

Fact: Rainy season begins June 15.

Data: Luxury Landscape will require three days to resod at bid of $4839. Guarantee includes six inspections in four-month period with necessary sod replacement at no extra charge.

Landscaping Professionals will require two days to resod at a bid of $3984. Guarantee includes six inspections in six-month period with necessary sod replacement at no extra charge.

Green Company will require three days to resod at a bid of $3707. No guarantee offered.

K-Mart Professional Crew will require three days to resod at a bid of $4000. Guarantee includes six inspections in 12-month period with necessary sod replacement at no extra charge.

Criteria: Twenty-square-foot area needs resodding. Budget allows $4500.00 expenditure. Guarantee required.

WRITING OPTIONS

1. *Incident reports.* Write an incident report on a real or imaginary business/industrial accident for your "employer." Suitable subjects are damaged equipment, brief fire, broken merchandise, minor burns or sprains, collapsed shelving, broken windows or doors, and so forth. Write a concise description of the incident or accident. Next, write a sequential analysis of the cause. Follow this with a review of the results, and, finally, present your recommendations to prevent the incident from recurring. Include graphics of the location and cost breakdowns.
2. *Periodic or progress reports.* Select one option.
   - **a.** In letter form write a periodic report on your monthly expenses to your parents or spouse. Include a circle graph or bar chart on the percentage on the actual dollar expense of each category. Include, or comment on the lack of, the following categories:

     Housing
     Utilities
     Food
     Transportation
     Clothes
     School Supplies
     Insurance
     Charge Account or Car Payments
     Grooming Aids
     Gifts or Miscellaneous

     Include other categories as they are appropriate to your expenses. Conclude with an appraisal of your expenses.
   - **b.** In memo form write a progress report to your academic advisor showing your progress towards completing a degree or receiving

a certificate or license. Begin with a statement of your overall goal and its requirements. Include tables of your courses, credits, grades, and grade point averages for courses completed, courses in progress, and courses remaining. Conclude with a discussion of your career plans.

c. Write a progress report on a project in which you are involved either in school or on-the-job. Include an introduction and sections on completed work, current work, future work, and an appraisal of progress. Be alert to graphic possibilities.

3. *Feasibility reports*. Select one option.

a. Write a feasibility report which analyzes a new purchase. State your purpose and the requirements. Then describe two or more probable alternatives (cars, office equipment, water beds, appliances, and so forth). Next explain a method for evaluating the products (survey, testing, research). Present your findings. Evaluate the data, and conclude with logical recommendations for purchasing one of the optional items.

b. Write a feasibility report which analyzes a procedure or problem in your place of employment. Identify the problem (the present means of advertising a product, scheduling personnel, awarding salary increases, providing in-service training, promoting personnel, handling tasks, or other similar procedures). State the standards which should be set to remedy the problem. Devise two solutions to the problem and analyze the merits and limitations of each. Interpret the feasible solutions to draw logical recommendations for implementing one or the other.

4. *Lab or Test Reports*. Select two simple products (ballpoint pens, glues, paints, stepladders, brooms, car waxes, toothbrushes, garlic presses, and so forth). With the purpose of determining which is the better product, devise a method or procedure to test each. Carry out your testing and then write a test report which states the object or purpose of the experiment, the explanation of the test method, a step-by-step analysis of the test and the results, and your conclusions and recommendations on which is the better product. The results section should offer the opportunity to present data in a table or performance curve.

5. *Field reports*. On your campus or at your place of employment select a location to conduct a field examination which will include conclusions and recommendations about the efficiency or safety of the location. Suggested fields to investigate are

On the campus	***At work***
parking lot layout	room furnishings
cafeteria food arrangements	office layout

classroom layout	restroom facilities
registration procedures	lounge or coffee room
recreational areas	locker space
library study carel layout	emergency exit doors/ stairwells
campus bookstore displays	storeroom arrangements
a piece of equipment	a piece of equipment
a small structure	security arrangements

Write a field report structured to include the purpose of your inspection, the methods for gathering data, the facts and results of your investigation, and the conclusions and/or recommendations which will make the chosen field more efficient or safer. Use appropriate headings.

6

LONGER REPORTS, PROPOSALS

Skills:

After studying this chapter, you should be able to

1. Identify the special features of longer reports.
2. Understand the terms *internal, external, solicited,* and *unsolicited* as they refer to proposals.
3. Devise appropriate streamlined or traditional formats for proposals.
4. Appreciate the purpose of transmittal correspondence and abstracts.
5. Write transmittal correspondence and abstracts.
6. Design title pages, table of contents, and lists of illustrations.
7. Attach supplements, appendices, or exhibits appropriately.
8. Write a lengthy proposal with all appropriate features.

INTRODUCTION

Although the need for conciseness is always present, a lengthy report is often necessary. In order to avoid "grey material," pages of dull grey type, a longer report is distinguished by special features to make the information contained in it more accessible. These features may include:

- Title Page
- Transmittal Correspondence
- Table of Contents
- List of Illustrations
- Abstract
- Supplements

Long-range planning programs, annual reports, or proposals are types of reports which may be so lengthy that they call for all of these features suitably presented in the bound folder. This chapter will cover the special features of longer reports by reviewing the organization and elements of a proposal.

PROPOSALS

A proposal is an action-oriented report. While most reports include recommendations for ongoing accomplishments, a proposal suggests a future task and includes a complete plan of how to accomplish this task. That is, a proposal contains procedure or equipment analysis, cost analysis, the capabilities of existing facilities, information on involved personnel, and usually a timetable for accomplishing the work. A report may recommend that a new policy be devised. A proposal suggests exactly what the policy must cover, a schedule for adoption, a procedure for implementing the policy, and the personnel who should be in charge. A proposal's purpose is to persuade the reader.

Types of Proposals

Proposals are classified as *internal* and *external, solicited* and *unsolicited.* Government and industry often solicit external proposals from outside agencies to solve problems or to develop services prior to awarding contracts or grants. A county commission may advertise for competing firms to submit proposals to develop a county-wide transportation system. A national airline may solicit proposals to develop a larger and faster jet. The Federal government solicits grant proposals for services in education, environment, energy, science, medicine, rural development, and other areas. A university may hire a consulting firm to develop a long-range expansion program. The agency which solicits external proposals spells out general requirements, cost ceilings, deadlines, and criteria for evaluation.

In business and industry the internal proposal, one written by a member of the organization, may be solicited or unsolicited. Management may appoint an individual or a committee to devise a program to change or improve some existing procedure or practice. A cafeteria manager may ask employees to submit proposals to increase sales in the fast-food line. An office manager may solicit proposals for policy and procedure to avoid charges of sexual harassment. Or any employee or group of employees may initiate an unsolicited proposal to management to purchase new equipment, improve working schedules, or alter procedures.

Writing a good unsolicited proposal for your employer is an excellent way not only to improve working conditions, but also to demonstrate your interest in and commitment to your company.

Organization and Format

A proposal explains an existing problem and proposes the concrete measures, procedures, or steps for rectification, along with an explanation of costs, equipment, personnel needs, and a time schedule. A proposal usually involves

- A clear statement of what is being proposed and why
- An explanation of the background or problem
- A presentation of the actual proposal, including methods, costs, personnel, and action schedules
- A discussion of the advantages and disadvantages
- The conclusions, recommendations, or an action schedule

All formal reports are more readable if they contain headings. The following topical headings should be considered for organizational purposes for a proposal although each actual proposal will suggest additional major and minor headings.

Streamlined format	*Traditional format*
Subject	Introduction
Objective	Purpose
Problem	Scope
Proposal	Background
Advantages	Investigative procedure
Disadvantages	Findings
Action	Proposal
	Equipment
	Capabilities
	Costs
	Personnel
	Timetable
	Consequences
	Advantages
	Disadvantages
	Conclusion

Writing the Proposal

For purposes of instruction we will concentrate on an unsolicited proposal, the kind you may originate in an entry-level career position or devise for consideration at your college or university. It is important to remember your audience. You essentially are writing a persuasive report; therefore, you must justify your proposal by presenting compelling reasons for its adoption. Although you may address your proposal to your immediate supervisor, a proposal is often reviewed by superiors further up the chain of authority. Your explanations, data, and language must be clear to those people who may not have any familiarity with the situation to which you address yourself. You must be objective and diplomatic.

Introduction. The introduction usually includes a statement of *purpose* and comments on the *scope* of the report or states the *subject* and *objective*. Write exactly what you propose along with a general statement of why the proposal should be given serious consideration. A statement of scope, if used, will orient your reader(s) to the material to follow. Examples are

> This report proposes the purchase of a table saw to increase production in our store fixture manufacturing plant and to increase profits. This proposal will document the problem, examine the capabilities of the proposed equipment, detail the costs, recommend the location, discuss the advantages, and present conclusions.

Purpose:	This proposal, to purchase and install bicycle supports and gate locks in the ABC Elementary School bike compound, is designed to eliminate vandalism and to decrease personal injuries.
Scope:	In this report an examination of the problem, the description of the proposed concrete bicycle supports and gate locks are presented, followed by the layout, costs, availability, and conclusions.
	or
Subject:	The establishment of a Standing Committee on the Status of Women.
Objective:	To explore and achieve equal employment and educational opportunities for women and to ensure the implementation of affirmative action and institutional change for women.

Problem/Background. The Problem/Background section details the existing problem, such as high costs, inefficiency, dangers or abuses, or low morale among employees. The solution you intend to propose will probably cost money or involve personnel in new responsibilities, so you must spell out that a very real and perhaps costly problem presently does exist. If it is not obvious how you researched the problem, you may need to include an explanation of your investigation techniques. If appropriate, research the operational costs of the present system. Project these costs over a week, a month, a year, or other appropriate time frames. Present data in tabular form.

Your problem may be that hazards or inconveniences exist under the present system. Document accidents, work slowdowns, late production schedules, or other related evidence. If the problem is causing low morale, research the turnover rate of personnel or incidents of friction.

You will realize that the background or problem material may be presented *inductively* by citing particular facts or individual cases followed by a general conclusion, or *deductively* by citing a general conclusion followed by the supporting facts or individual cases. Modern business writing leans more and more to the deductive method without, however, entirely abandoning the other.

Compare the following and ponder which would be more easily read and understood by someone unfamiliar with the existing problem, a decline in the company's sales volume:

Draft 1 (Inductive)

Gross margins declined from 13.8% in the first quarter of 1982 to 13.1% in the corresponding 1983 quarter. There was a drop of 26% in sales volume between these two periods. At the branch level new unit margins dropped from 15.8% to 12.2%, but 50% of all units sold in the first quarter of 1983 were from inventory made prior to 1983 . . . and so on to an eventual statement that all this has reduced profits.

Draft 2 (Deductive)

Company profits have been adversely affected by a decline in sales volume accompanied by a drop in gross margin. Sales in the first quarter of 1983 were 26% below the volume in first quarter 1982. In the same period gross margin decreased from 13.8% to 13.1%. The principal reason for the lower gross margin in 1983 was the fact that more sales were from inventory rather than new production.

The first draft loses the reader in a mass of figures while the second draft presents the key thought in the first sentence. It helps the reader to understand the statistics.

In a proposal to purchase a computer for two floral shops the problem was documented as follows:

Problem:

The testing problem is that bookkeeper transportation and telephone costs are excessively high. Because both of our floral shops maintain separate inventory control of customer account information, Bookkeeper Mary Jacobs must frequently travel between the two shops and telephone for customer account information. In addition, this system requires twice the amount of time necessary to review the total daily sales. Table 1 shows the transportation and communication expenses of the bookkeeper under the present system:

Table 1 Current Bookkeeper Transportation and Telephone Expenses

Time Frame	Travel Time (hr)	Gas ($)	Telephone Time ($)	Cost @ $6.00 per hr wage
Day	1	5	1	6
Week	5	25	5	30
Month	20	100	20	120
6 Mo	120	600	120	720
Year	240	1200	240	1440
		TOTAL COST $2640.00		

The $2640.00 wage and gas expenditure can be put to more productive use.

The emphasis here is on inefficient costs. The following is from a proposal to establish a regulating committee to end sex discrimination at a community college. The problem does not entail costs, but concrete evidence:

Problem:

Some sex discrimination facts present themselves:

1. Of the 215 faculty members 79 or 36.7% are women, a percentage which is not reflected in either administrative positions or standing committee membership.
2. Of the 36 administrators only three (3) are women; of the ten (10) division chairpersons none are women; of the twelve (12) department heads only two (2) are women; of the twenty (20) area leaders only five (5) are women; of the total 78 positions only ten (10) or 12.8% are women.
3. Of the 207 faculty and administrators serving on standing committees only 52 or 25% are women. One committee has no women members.
4. Women have voiced concern that the inequitable number of women in administrative and standing committee positions is a "negative incentive" for innovative teaching, volunteer assignments, and request for advancement consideration.

5. Women students have voiced concern through the agency of the student government association that the College does not actively counsel and provide for women students to excell nor to set goals commensurate with the expanding opportunities for entry into the previously male-dominated professions.

Be thorough and exacting in your documentation of the problem. The information in this section will be referred to when you detail the consequences of adopting your proposal.

Proposal. Present the solution to the problem by providing all of the particulars of your proposal. As already suggested you may wish to consider the following subheadings:

Equipment/Procedure

Capabilities

Costs

Personnel

Timetable

If the proposal involves the purchase of new equipment, describe it accurately and explain its function and capabilities. If the proposal involves a new procedure or policy, explain exactly how it will work. Under costs include initial purchase price, financing, installation, labor, and training costs. If new or transferred personnel are involved, include the qualifications for the position or the qualifications of the employee. Summarize duties and salaries. If you use the traditional format, include a timetable or work plan for making your proposal operable in this section. If you use the streamlined format, place your schedule of tasks to be accomplished in the final Action section. An effective proposal not only seeks action but also becomes the blueprint by which the action shall be performed. Because the actual proposal section of your report is often lengthy, samples are not given here but may be reviewed in Figures 6.1 and 6.2.

Consequences. This section may be divided into *Advantages* and *Disadvantages* or *Strengths* and *Limitations*. Here you refer to the details in your Problem section and explain how each problem will be solved or alleviated.

If your proposal costs money, show in graphics how implementation will save the company money over projected periods of time. Often a costly expenditure will ultimately save the company money over a few years due to increased efficiency or production. If you propose a new system, number and list all of the benefits of it. Employee morale is an important concern of management; if adoption of your proposal will improve employee morale, detail these benefits.

Do not ignore disadvantages (temporary work stoppage, layoffs of employees, limited capabilities, and so forth). Discuss these disadvantages or limitations in a positive manner.

Following are brief sections from proposals outlining Consequence. The first example is from a proposal to pave an American Legion Hall parking lot:

Consequences:

By paving the existing parking lot we will be able to accommodate 30 or more cars than the present 140. By eliminating the problems of dust and erosion, the physical appearance and value of the property will be enhanced. During the two weeks of construction Sunset City officials have agreed in writing to allow our 57 employees and approximately 100 daytime visitors to park in the Sunset City Hall west parking lot.

The second example is from a proposal to adopt new displacement and furlough rules for an airline pilots' association:

Advantages:

Adoption of the system-wide seniority rules in the pilot displacement and furlough process results in the following advantages:

- Senior pilots are able to displace at any base.
- The total number of displacements arising out of a curtailment situation is small, and the displacements are mostly confined to the base where the curtailment situation occurs. Thereafter, significantly fewer pilots are affected by curtailment.
- A maximum of two base moves exists for every pilot curtailed.
- Protection of captaincy is guaranteed for all but the junior captain at base. The junior captain has base protection.
- The indicated cost savings is 30 percent over the current operating agreement.

Disadvantages:

These benefits are achieved at the cost of limiting pilots' freedom of choice in the following manner:

- A pilot's displacement choices are limited.
- If a captain desires to revert to first officer status to displace in a particular equipment, he may not be able to do so.

Conclusions. The body of your proposal has examined the problem and presented a blueprint for action. You have anticipated the questions and possible objections to your proposal and objectively, but persuasively, responded to them. Your closing section is essentially an urge to action. In the traditional format summarize your proposal features and reemphasize the advantages in numbered statements. Close with a persuasive statement, such as

> I urge you to give serious consideration to this proposal and am available to discuss its particulars with you.

In the streamlined format, title your section *Action* and present your schedule, work plan, or timetable. Figures 6.1 and 6.2 illustrate the optional closing sections for proposals.

Your signature, position, and date are appended to the final page in the right-hand corner of the paper:

Janet Gorky 1/15/8X
Janet Gorky, Assembler 11/15/8X

PRESENTING THE LONGER REPORT

As previously mentioned a number of special features are included with a longer report. After preparation, all of the parts of the longer report are usually submitted in a bound folder.

Title Page

Include an attractive and clarifying title page. This page should include

- A precise title
- The name, title, and company of the person(s) to whom the report is directed
- The name, title, and company of the writer(s)
- The date

A precise title, such as

> PROPOSAL FOR PURCHASE AND INSTALLATION
> OF IONIZATION AND PHOTOELECTRIC FIRE
> ALARM SYSTEMS IN OCEANVIEW CONDOMINIUM UNITS

is more effective than a vague title, such as

PROPOSAL TO DECREASE FIRE HAZARDS

Figures 6.1 and 6.2 include title pages to longer reports.

Transmittal Correspondence

A letter or memorandum of transmittal accompanies most longer reports. The purpose of the transmittal correspondence is to convey the long report or proposal in a suitable explanatory manner. It is usually very brief, three or four short paragraphs. It contains

1. The title and purpose of the report
2. A statement of when it was requested or why it is being submitted
3. Comment on any problems encountered
4. Acknowledgement of other people who assisted in assembling the report

Do not include repetitions of the data in the actual report. Figures 6.1 and 6.2 illustrate typical transmittal correspondence.

Table of Contents

Your reader(s) will want to be able to refer to sections quickly. A table of contents not only helps the reader(s) to turn rapidly to a particular section of the report, but also gives an initial indication of the organization, content, and emphasis of the report. A table of contents should accompany every written report that exceeds eight or ten pages and may be helpful in some shorter reports. Title the page *Table of Contents*. All headings used in the report are included in the table, and the subordination of sections is indicated by indentation. The starting page of each section is included as shown in the sample proposals in Figures 6.1 and 6.2.

List of Illustrations

If graphics (tables and figures) are used throughout your report, include their table and figure numbers, titles, and page references on the same or separate page from the Table of Contents. Center the title *List of Illustrations*. List tables separately from figures. Figures 6.1 and 6.2 illustrate lists of illustrations.

Abstract

An abstract is a brief *informative* or *descriptive* summary of a longer report. It is written after you have completed the full report, but it is intended to be read first by your audience. Occasionally the abstract is placed in the transmittal correspondence, but modern usage calls for it to be presented on a separate page placed after the table of contents and list of illustrations.

The *descriptive abstract* only identifies the areas to be covered in the report. It serves as an extended statement of scope and is useful only for a very extensive report because it indicates the report's organization, but not its content.

An *informative abstract* summarizes the entire report allowing the reader an overview of the facts before proceeding to read all of the detail. An abstract is seldom longer than one page. It should never exceed 10 percent of the length of the entire report or it defeats its purpose.

Besides expecting an abstract to accompany a formal report, a busy executive often requests assistants to prepare abstracts of newspaper and periodical articles in order that the executive might have access to the content of pertinent reading material in brief form. He or she can then decide whether to read the entire report or article.

Librarians subscribe to abstracting journals which provide brief descriptive or informative abstracts of articles published in a number of journals in a specific field. These allow the reader to obtain an overview of content without lengthy reading. The reader can then reject or pursue the lengthy articles.

Many busy people are likely to review your formal report or proposal. Most will rely on the information in the abstract to make initial judgments. Because these readers may not possess the technical knowledge and language which your full report embraces, be careful to avoid technical terminology.

To write an informative abstract, follow these steps:

1. Read the entire report to grasp its full contents.
2. Estimate the number of words and plan an abstract which will be approximately 10 percent of the original length.
3. Write down or underline in a report copy the key facts, statistics, and points under each heading.
4. Do not include the statement of scope.
5. Omit or condense lengthy examples, tabulated material, and other supporting explanations.
6. Rewrite the information which you have culled from the report in original sentences.
7. Edit the abstract for completeness and accuracy.

Following is the text of a brief proposal. The key points that should be included in an abstract have been underlined.

PROPOSAL TO HIRE PART-TIME
TELEPHONE RECEPTIONIST

INTRODUCTION

Key Purpose

Purpose:

The purpose of this proposal to hire a part-time telephone receptionist is to increase the work productivity of the present employees of Bruce Essex Supply Company.

Scope:

This report will document the problem, propose the hiring of the telephone receptionist, detail the cost, duties, hours, and location, and review the advantages.

PROBLEM

Key Problem

Presently, the pricing clerk is assigned telephone reception duties. This situation is costly to the company because the pricing clerk is salaried in excess of the demand for such duties, and these duties interrupt the normal productivity of the pricing clerk. Further, due to the volume of phone calls, the pricing clerk must both work overtime and leave incomplete work.

Key Cost Factor

The present system costs the company $7440.00 per year as shown in Table 1:

TABLE 1 COMPANY COST FOR TELEPHONE RECEPTION

Time	Calls	Time per Call at 45 sec	Cost at $5.00 per hour ($)	Estimated Overtime (hr)	Overtime at $7.50 per hour ($)
Day	400	300 min	25.00	1	7.50
Week	2000	25 hr	125.00	4	30.00
Month	1000	100 hr	500.00	16	120.00
6 Mo	48000	600 hr	3000.00	96	720.00
Year	96000	1200 hr	6000.00	192	1440.00

Total Cost of Telephone Duties: $7440.00 per year

Even though the pricing clerk is paid overtime, she is frustrated, overworked, and operating inefficiently.

PROPOSAL

Therefore, I propose the hiring of one part-time telephone receptionist to work six hours a day, thirty hours per week.

Key Cost Factor

Cost:

Total annual wages will be $4824.00 as shown in Table 2:

TABLE 2 COST FOR PART-TIME TELEPHONE RECEPTIONIST

Time	# of Hours	Cost at $3.35 per hour ($)
Day	6	20.10
Week	30	100.50
Month	120	402.00
6 Mo	720	2412.00
Year	1440	4824.00

The $3.35 per hour wage is in compliance with the 1981 minimum wage law.

Hours:

Because the majority of phone calls are between 9:00 a.m. and 4:00 p.m., recommended hours for employment are from 9:30 a.m. to 3:30 p.m. The telephone receptionist would have the opportunity for one or two hours extra work at the same wage during sales if management feels it is necessary.

Key Duties

Duties:

Duties of the telephone receptionist should include transferring calls to the proper departments, screening management's incoming calls, light filing, typing, and posting the mail toward the end of each day.

Location:

The telephone receptionist may be positioned at either of the two extra desks which are already equipped with office phones. The existing switchboard can be moved from the pricing clerk's desk to the selected desk.

Schedule:
I propose that the operator be hired by April 15, 198X.

ADVANTAGES

This proposal has numerous advantages:

Key Cost Factor

1. The company will save $2616.00 on the cost of telephone answering services.

Persuasive Factor

2. Because the position will be part-time, the company will not have to offer fringe benefits, such as overtime pay, profit sharing, or insurance.

3. By allowing the pricing clerk to concentrate on her duties, the position will be more cost efficient.

CONCLUSION

My proposal, if adopted, should prove to be both economical and practical. I urge you to give it serious consideration.

The proposal consists of approximately 500 words. There are definitely other important factors besides those that are underlined, but to include them would entail too lengthy an abstract. The following abstract for the preceding proposal consists of 59 words.

> I propose we hire a part-time telephone receptionist. Presently the pricing clerk assumes the telephone duties at a company cost of $7440 per year. A part-time operator will cost $4824. She will handle calls and perform other light duties. The company will save $2616 on the cost of telephone answering services, avoid fringe benefits, and realize more efficiency from the pricing clerk.

An abstract summarizes the concrete data. Additional sample abstracts appear in Figures 6.1 and 6.2.

Supplements

It may be helpful to attach various supplemental materials to your proposal. The body of your proposal may have summarized complex financial projections, tabulated the capabilities of a piece of equipment, or recommended qualified personnel to assume new responsibilities. Rather than include all of the figures, specifications of equipment, or personnel resumes in the report, these supporting materials may be attached as exhibits or appendixes and grouped in the back of the binder. Should you include supplemental materials, follow these guidelines:

- Number exhibits and letter appendixes; title each.

Examples:	Exhibit 1	Interview Topic Outline
	Exhibit 2	Sample Policy Matrix
	Appendix A	Forecast of Operation and Return on Investment
	Appendix B	Resume on Manager Donald C. French

- Refer to the exhibits or appendixes in the body of your proposal.
- Include the exhibit or appendix titles in your table of contents.

Figures 6.1 and 6.2 include supplemental materials following the Conclusion sections of the proposals. Note their listings in the appropriate tables of contents and their references in the texts. Figure 6.1 does not include the supplemental material in this text.

PROPOSAL TO PURCHASE
ROLLAWAY FILES
FOR AERIAL PHOTOGRAPHS

Prepared for

John Smith
Assistant Director
Ace Aerial Photography
Fort Lauderdale, Florida

by

Mary Jones
Laboratory Technician

January 10, 198X

Figure 6.1 Sample streamlined proposal with formal report features

MEMORANDUM

TO: John Smith, President

FROM: Mary Jones, Laboratory Technician M.J.

SUBJECT: Proposal to Purchase Rollaway File Unit

DATE: January 10, 198X

Enclosed is a proposal to purchase two, 2-drawer, Esselte rollaway file units for aerial photographs. Adoption of the proposal will economize operating expenses and increase office efficiency.

Ms. Juanita Colby, purchasing department, was instrumental in obtaining the costs of available rollaway files.

May I hear from you about this proposal by Friday?

MJ:eg

Enclosure

ii

Figure 6.1 Continued

TABLE OF CONTENTS

LIST OF TABLES

iii

Figure 6.1 Continued

ABSTRACT

The present system of storing aerial photographs is inefficient and costly. The laboratory technicians must locate photographs for realtors, engineers, and other clients who desire dated copies and glossies. The Company is spending $2160.00 a year on the search procedure, and the searches are disrupting the efficiency of the laboratory technicians. I propose the purchase of two 24″ Esselte Tu-Dror rollaway file units for $462.00. Required search time will be reduced by 324 hours per year. The file unit system will save the Company $1586.00 a year following initial purchase, and $1940.00 each year thereafter.

iv

Figure 6.1 Continued

1

SUBJECT

The purchase of two rollaway files for aerial photographs.

OBJECTIVE

To improve search efficiency and to resolve operation costs.

PROBLEM

Presently, the aerial photographs are stored on the second floor in the east wing in boxes in no sequential order. When an engineer, realtor, or other client requires an aerial photograph on file, a technician must search through the boxes to locate the photograph. Table 1 shows the time and wages expended in searches under the present system:

TABLE 1
TIME AND WAGES FOR SEARCHES OF AERIAL PHOTOGRAPHS

Time Frame	# of Searches Conducted	Required Search Time (30 min ea)	Cost of Searches $6.00 per hr wage ($)
Daily	3	1 hr 30 min	9.00
Weekly	15	9 hr	54.00
Monthly	60	30 hr	180.00
Quarterly	180	90 hr	540.00
Yearly	720	360 hr	2160.00

The department is expending $2160.00 a year for employee time to search for aerial photographs. In addition to the unnecessary cost, employees are frustrated by the inefficient storage, and clients are impatient.

Figure 6.1 Continued

2

PROPOSAL

Therefore, I propose the purchase of two rollaway file units. Table 2 shows the costs and features of three types of rollaway files:

TABLE 2
COMPARATIVE ROLLAWAY FILE UNITS

Esselte Model	Features	Dimensions H	Dimensions W	Dimensions D	Cost ($)
Pendaflex	24″ desk side unit Sliding cover Steel storage shelf Rubber ball bearing casters	28½	13	24¼	107.00
Tu-Dror	24″ two drawer unit Suspension bottom drawer Sliding cover with lock Rubber ball bearing casters Locking brake	30¼	15	24	231.00
Junior Pendaflexer	18″ desk side unit Sliding cover with lock Steel storage shelf Rubber ball bearing casters	28½	13	18¼	99.25

I recommend the Tu-Dror units at $231.00 each or $462.00 for two. Four drawers are necessary to file existing photographs and to allow space for the estimated volume of 198X photographs.

Figure 6.1 Continued

3

ADVANTAGES

1. Purchase of the file units will result in a cost reduction of $1486.00 for the first year and $1948.00 thereafter. Table 3 shows the time and wages involved in searches using the file unit system:

TABLE 3
TIME AND WAGES FOR SEARCHES — NEW SYSTEM

Time Frame	# of Searches conducted	Required Search Time (3 min ea)	Cost of Searches ($6.00 per hour wage)
Daily	3	9 min	.90
Weekly	15	45 min	4.50
Monthly	60	3 hr	18.00
Quarterly	180	9 hr	54.00
Yearly	720	36 hr	212.00

2. Clients will be satisfied that it takes the technician less time to search for aerial photographs which will be in coded order.
3. The space that is used for aerial photograph boxes will be available for other supplies.

ACTION

1. The files should be purchased by February 1, 198X.

2. The three laboratory technicians will require one month for coding and filing existing photographs. Work is generally slow in February.

3. I will assume the authority for purchase order and directing file duties.

Mary Jones 1/10/8X

Mary Jones 1/10/8X
Laboratory Technician

Figure 6.1 Continued

PROPOSAL TO DEVELOP
A REIMBURSABLE EXPENSE POLICY
AND
MONITORING AND CONTROL SYSTEM

Prepared for
Mr. Duane P. Morton
President
Ace Electronics Company, Inc.
Fort Lauderdale, Florida

by

Charles E. Smith
Senior Partner
Smith Business Consultants
Fort Lauderdale, Florida

February 24, 198X

Figure 6.2 Sample traditional proposal with formal report features (Courtesy of Charles E. Smith; adaptation)

SMITH BUSINESS CONSULTANTS
1000 Plaza Court
Fort Lauderdale, Florida

February 24, 198X

Mr. Duane P. Morton, President
Ace Electronics Company, Inc.
2120 West Broward Boulevard
Fort Lauderdale, Florida 33316

Dear Mr. Morton:

We have completed the study which you authorized in November 1980 concerning the control of reimbursable business expenses at Ace Electronics Company and are pleased to submit the enclosed proposal to make expense practices more equitable.

We appreciate the cooperation of your officers and other employees in conducting our study. All transcribed interviews and records research findings will be kept in confidence for six months should you need to review this material.

We thank you for this opportunity to be of assistance.

Very truly yours,

Charles E. Smith

Charles E. Smith
Senior Partner

CES:jsv

Enclosure

ii

Figure 6.2 Continued

TABLE OF CONTENTS

iii

Figure 6.2 Continued

ABSTRACT

This report for Ace Electronics Company, Inc. proposes the development of a reimbursable expense policy along with a monitoring and control system to make expense practices more equitable, to allow for abuse detection, and to reduce costs.

Extensive interviews and expense records research reveal that expenses are excessive, existing policy is inadequate, and methods of approval and control are deficient. By implementing this proposal the approximate yearly $3.3 million reimbursable expenses may be reduced by as much as $1 million.

We propose the development of a written reimbursable expense policy which incorporates an expense matrix of all allowable expenses, categorizes employees, sets conditions and restrictions, and clearly states approval individuals. Further, we propose procedural regulations which establish periodic report periods and deadlines, entail new report forms, and institute a definite system of review and analysis. A work plan to carry out these proposals includes the appointment of a director, the establishment of committees, and a listing of chronological duties.

By implementing this proposal reimbursable expenses shall be reduced, and adoption may well lead to cost reduction sensitivity in other expense areas.

iv

Figure 6.2 Continued

PROPOSAL TO DEVELOP A REIMBURSABLE EXPENSE POLICY
AND MONITORING AND CONTROL SYSTEM

INTRODUCTION

Purpose

This proposal to develop a reimbursable expense policy and monitoring and control system is designed to make expense practices more equitable, to establish a system which will detect and prevent abuses, and to reduce costs.

Scope

This report presents a description of our study, a review of our findings, a dual proposal, the advantages of implementation, and the conclusions.

BACKGROUND

Procedure

To determine the facts relating to reimbursable expenses we used a sample basis. With the assistance of four senior executives, a sample of 25 of the approximately 100 executives at Ace was chosen. Personal interviews were held with twenty of them, and extensive records research was conducted for fifteen covering the second quarter of 198X, the period chosen for the analysis.

Our first step was to interview these officers, two financial managers, and the personnel manager and to examine a sampling of expense reports. From that work we prepared two documents: a list of discussion questions for the interviews, and an outline of the records research requirement. Appendixes A and B exhibit these documents and indicate the extensive nature of the fact finding.

The Smith consultants conducted the interviews, and an Ace team directed by the assistant corporate controller conducted the records research.

1

Figure 6.2 Continued

2

Findings

A. Scope of Expenses

Reimbursable business expenses represent a significant controllable cost at Ace. During the first nine months of 198X total expenses for officers and others was $2.5 million, an annual expense of over $3.3 million a year. For the officer group alone the figure was $1 million a year.

B. Policies and Practices

Numerous examples of excessive expense practices and failure to control expenses are evident upon examination of the records and by admission of the officers in confidential interviews. We believe that reimbursable expenses at Ace could be reduced by $1 million a year without adversely affecting the business.

As a result of unclear or inadequate policy or poor communication, most interviewers displayed uncertainty about what was expected of them. During the interview officers were asked: "Can you identify the written expense policy/procedure as it relates to you?" Following are representative answers:

"One might exist, but I'm not familiar with it."
"There's only a memo on transportation."
"There is no written policy."
"Policies are not spelled out. I'm not familiar with any written policy. I do what I did for my former employer."

Further, existing written expense policies lack clarity.

- A limited number of items that may be authorized are listed, but the conditions under which they are authorized are not always clear. For example, statements, such as "when time is a factor," "where necessary for company business," and "only to employees who hold the types of jobs that require such meetings," appear in the policies.
- Some expenses require prior approval, but the approving individual is not always identified.
- Some items of allowable expenses, such as home entertainment and telephone answering services, are not treated.

Figure 6.2 Continued

3

- The frequency of reporting is not covered.
- A few items, such as social club memberships and access to the executive dining room, are inadequately treated. In some cases these are considered executive perquisites and in others reimbursable expenses.

C. Approval and Control

Methods of approval and control are inadequate.

- Officers report they are not comfortable when questioning the expense practices of close associates, and rarely do so.
- Expense documentation and explanation are inadequate. Reporting forms fail to determine such things as class of air travel, cost of overnight accommodations per night, purposes of business meetings, the number of people entertained, and so forth.
- Budgetary control is difficult. Monthly reports do not show expenses by individuals, but by divisions.
- Expense reporting is often tardy. For the sample officers timeliness ran from prompt (2 to 3 days after the reported period) to late (6 months). Many officers reported weekly, others monthly, and some on less periodic schedule.
- The expense report is not a complete record of expenses. Some officers are reimbursed for expenses directly from petty cash, or the invoice for an expense item is paid directly by Ace and not shown on an expense report.

D. Variations

The interviews and the records research reveal a wide range of variable practices.

- In the sample of the fifteen officers whose expense reports were analyzed, only 7 charged telephone expenses, 11 had social club expenses, 10 were reimbursed for gifts, and 5 incurred expenses for personal entertainment.
- Air travel practices further illustrate the variety that exists. During the analyzed quarter 3 of the officers made at least one flight, but 2 of them did not submit ticket receipts. Of the 11 other employees whose flights could be analyzed, 6 flew first class and

Figure 6.2 Continued

4

only 2 paid the difference between first class and coach fare. Seven wives flew with their husbands at company expense, 2 of them first class.

- There is considerable variation in the cost of overnight room accommodations. The range is from $26.00 to $110.00 a night.
- The average per meal cost of business meetings range from $10 to $48. Business meetings were usually conducted at lunch.

The great variation in expense practice is the basis for our conclusion that substantial opportunities for cost reduction of at least $1 million a year exist.

PROPOSAL

To correct these deficiencies in the reimbursable/expense system, we propose

1. the development of a written reimbursable expense policy, and
2. the establishment of firm procedural regulations.

Expense Policy

Ace should prepare a written reimbursable expense policy that is clear with regard to each type of expense, that accommodates differences among employees, and that covers certain procedural requirements. To accomplish this

1. All personnel should be divided into categories. All employees should not be treated uniformly. Their expense spending requirements and privileges should be recognized in the policy. We recommend that the two categories be designated as follows:

Members of the Board Vice-presidents	All Other Employees

2. A policy matrix should be devised which lists all reimbursable expenses, employee category differences, restrictions and conditions, and approval authority. Appendix C illustrates a sample policy matrix.

Figure 6.2 Continued

5

3. A clarifying policy should be written for wide distribution and inclusion in the Administrative Policy Manual. The policy should include guidelines and standards, procedures for advances, procedures for preparation, processing, and approval of the expense report, and specific guidelines and standards for each type of reimbursable expense in the marix. Appendix D shows a sample administrative policy page.

Monitoring and Control

Procedural regulations should be established.

1. The report period should be monthly; deadline for submission should be one week thereafter.
2. Expense report forms should be revised to require documentation and explanation.
3. The Expense Report must be established as the sole vehicle for recording and reimbursing expenses.
4. Responsibilities of the reporting individual, the controller's department, and the approving individual should be developed into a definite system.
5. The role of the controller's department should be expanded. An editing function there should first evaluate the adequacy of the documentation and explanation, and the Expense Report should be returned to the reporting individual for correction if there are deficiencies. Next the report should be examined for conformance to policy, and any exception should be noted on a Buck Slip which, along with the Expense Report, should be forwarded to the approving individual.

   The controller's department should prepare a summary analysis of each individual's expenses on a quarterly basis. Quarterly and cumulative expenses should be compared to budget.

Figure 6.2 Continued

6

Work Plan

I. Staffing

A. Appoint the Vice-president of Finance as Project Director with responsibility for implementing all recommendations.

B. Appoint a Corporate Expense Policy Committee to work with the Project Director.

C. Assign a small staff to work with the Committee without interruption.

II. Expense Policy

A. Establish the basis for assigning all personnel into the recommened two categories.

B. Prepare an expense matrix and policy for each type of reimbursable expense for each category of personnel.

C. Submit the policy to officers for discussion and modification.

D. Approve the policy.

E. Disseminate the policy and conduct familiarization training.

III. Monitoring and Control

A. Revise Expense Report format and create Daily Expense Diary, Buck Slip, and Quarterly Analysis.

B. Establish the edit and analysis functions in each division

C. Train expense report editors and analysts.

D. Test the system and make adjustments.

E. Commence live operation of the system.

ADVANTAGES

By developing a written reimbursable expense policy and establishing firm procedural regulations, the following advantages should accrue.

1. Reimbursable expense practices should become more equitable over the broad organization due to clarification and control.
2. A system for detection and prevention of abuses will be established.
3. Superiors shall exercise more control of expenses.

Figure 6.2 Continued

7

4. Quarterly analysis should encourage discipline and facilitate budget preparation.
5. Implementation should reduce costs.

CONCLUSION

An attractive scope for cost reduction exists and provides the basis for achieving improvements through implementing innovations in policy and control. In our opinion, the recommendations in this report are well worth implementing and, in addition, will lead to sensitivity towards cost reduction in other areas.

Charles E. Smith 2/24/8X

Charles E. Smith 2/24/8X
Senior Partner

CES:jsv

Figure 6.2 Continued

APPENDIX A

REIMBURSABLE EXPENSES INTERVIEW TOPICS

1. Position held during second quarter, 198x
   a. Title
   b. Superior, subordinates
   c. Nature of position
   d. Approval authority for expense reports
2. Whose expense reports do you approve?
   a. Are summaries prepared?
   b. How do you control subordinates' expenses?
   c. Do you ever question their expenses? Details.
   d. Have you ever had a drive to reduce expenses? Explain.
3. Who approves your expense reports?
   a. Are summaries prepared?
   b. Are your expenses ever questioned? Details.
   c. Do you ever seek prior approval for expenses? Details.
   d. Do you have any prior understanding with regard to your expenses?
4. What is the budgetary control over your expenses?
   a. In what account in what cost center are your expenses accumulated?
   b. How do your expenses compare to budget?
5. Have you ever been told what the expense policy is as it relates to you? Details.
6. Can you identify the written expense policy/procedure as it relates to you? Details. What are the salient points?
7. Identify the forms that are used: expense reports, petty cash disbursements, report of outstanding advances, summaries.
8. What is the frequency of reporting and approval and what is your timeliness? Any difficulties?
9. What are the controlling policies/procedures with regard to advance accounts, and what are your practices?
10. List all of the categories of your reimbursable expenses.

8

Figure 6.2 Continued

9

11. Do you incur reimbursable expenses that are not shown on your expense report? If so, where are they recorded?
12. What are the significant restrictions on your expense spending practices?
13. Do you ever misrepresent expense items? e.g. to combine minor items, to recover excessive charges?
14. What is the influence of your status on your expense practices? Is there discrimination by rank? Amplify?
15. How do you view the company attitude concerning expenses? Have there been any changes? How does it compare with other companies in which you have worked?

Figure 6.2 Continued

APPENDIX B

RECORDS RESEARCH OUTLINE

I. Sources of information
   A. Expense report
   B. Summary business expense report
   C. Petty cash disbursement form
   D. Report of outstanding advances
   E. Approval authority
   F. Quarterly Budget Report for appropriate cost center
II. Information required
   A. Controlling policy/procedure
   B. Approval authority
   C. Summary
   D. Type of form
   E. Completeness and deficiencies
   F. Date and total expenses for period
   G. Documentation check
   H. Assumptions
III. Quarterly Summary
   A. Transportation
   B. Hotel rooms
   C. Meals
   D. Business meetings
   E. Other
IV. Analysis
   A. Personnel category differences
   B. Direct payments
   C. Nature and amount of petty cash disbursements
   D. Timeliness
   E. Discrepancies
   F. Significance

10

Figure 6.2 Continued

APPENDIX C

SAMPLE REIMBURSABLE EXPENSES POLICY MATRIX

Type of Expense	EMPLOYEE CATEGORY President Vice-Pres or above	All Other Employees	Restrictions & Conditions	Approval Required
Airline and Railroad	First Class Reimbursable	a) Coach reimbursable b) First Class Restricted	Restrictions: a) Traveling in company of Pres., V-P, or above b) Duration of trip exceeds 7 hr c) Traveling with customer, etc	President Vice-Pres or above
Buses	Reimbursable	Reimbursable		
Rented Cars	Reimbursable	Reimbursable	Condition: a) Other forms of public transport cannot meet business requirements b) Standard size low-priced car	
Personal Cars	Reimbursable	Reimbursable	Condition: Must not be used on long trips when other transport is less expensive and obtainable	
Taxi Cabs	Reimbursable	Reimbursable	Condition: Other forms of public transport cannot meet business requirements	
Limousines and Company Cars	Restrictively Reimbursable	Restrictively Reimbursable	Restriction: a) Essential b) Mileage log maintained	President Vice-Pres or above
Air Charter	Reimbursable	Reimbursable	Restriction: Extreme emergency situation only	President Vice-Pres or above
Lodgings Single Rooms First Class Accommodations	Reimbursable	Reimbursable		
Suites	Restrictively Reimbursable	Restrictively Reimbursable	Restriction: Essential for the conduct of business	President Vice-Pres or above

11

Figure 6.2 Continued

APPENDIX D

SAMPLE ADMINISTRATIVE POLICY PAGE

ACE Administrative Policy Manual			
Section	Dist. List	Date Issued	Policy No.
REIMBURSABLE BUSINESS EXPENSES	3	12/30/8X	RB-1
Subject			Page of
Ace Reimbursable Business Expense Policy			3 39

III. TYPES OF REIMBURSABLE BUSINESS EXPENSES

For ease of reference, a matrix summarizing reimbursable expenses has been included in Section XI of this Policy.

A. Transportation Expenses

All requests for transportation should be submitted through the Transportation Department.

Where use of the Transportation Department is impractical, an employee may arrange his own transportation.

The restrictions and documentation required for transportation expenses are as follows:

1. Airlines and Railroads - Coach type accommodations are to be used. All expenses in this category must be documented by a ticket stub when submitting the Ace Expense Report. When a ticket purchased by Ace is not used, it should be returned to the Controller's Department with the expense report for a refund. The amount of the unused ticket should be included on the expense report.

   When traveling to and from the airports and Ace, public transportation should be used wherever practical.

2. Buses - The use of long-distance buses is permitted when they can reasonably meet the needs of the Company.

12

Figure 6.2 Continued

APPENDIX D

3. Rented Cars - Use of rented cars should be limited to those situations where public transportation is not available or cannot meet business requirements. When the situation requires a rented car, short-term arrangements should be made by the individual. On rentals exceeding one month, arrangements should be made through the Ace Purchasing Department. On all rentals, a standard size, low-priced car should be used. In all cases optional insurance should not be purchased and where national car rental services are used, the discount should be obtained. When submitting this expense for reimbursement, a copy of the itemized invoice is required.

13

Figure 6.2 Continued

EXERCISES

1. Locate, photocopy, and abstract an article from a professional journal in your field. Submit the copy of the article along with your abstract. Remember that your abstract should summarize the key *facts* and should not exceed 10 percent of the report length.
2. Write an abstract for the following brief proposal.

Subject: Modification of the North Campus perimeter road and parking lot access road speed bumps.

Objective: To alleviate complaints and to prevent further damage to student and visitors' automobiles.

Problem: Since the installation of 80 speed bumps on the campus perimeter road and parking lot access roads in March 198X, over 100 students and other campus visitors have registered written complaints to the Security Department regarding damage to their automobiles. The height of the bumps is 8 inches; compact cars have only an 8-inch clearance. Further, 47 percent of the complaints are by drivers of large and intermediate automobiles. Reported damages include front end misalignments, rear end leakage, and muffler dents.

Inspection reveals that the bumps are excessively gouged and scraped. While some automobile damages may be due to excessive speed, a State Road Department inspector concurred, following his June 15, 198X, visit, that the bumps are too high and too abrupt. Appendix A exhibits a copy of his letter.

Proposal: Therefore, we propose that the 80 speed bumps be modified by grinding the peaks to a 6-inch height and sloping the sides with asphalt to a 35-degree slant. These modifications will eliminate the shock that automobiles are experiencing yet still deter speeding. Figure 1 shows a cross-section of the speed bumps before and after modification:

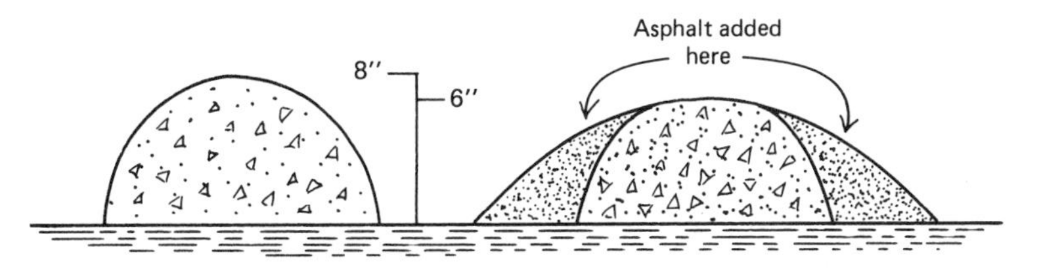

Figure 1 ***Cross-section of a typical speed bump before and after modification***

Equipment: Because the college does not own the equipment necessary for the modifications, the equipment must be rented. Necessary rental equipment includes

- one hand-held, gas-operated, heavy duty grinder with emery stone disc,
- one gas-operated hot asphalt mixer, and
- one water-filled, hand-operated 500 pound roller.

Labor: No special skills are needed to operate the equipment. Three Maintenance Department personnel can complete the modifications in an estimated 120 hours over two weeks.

Procedure: One man can grind the bumps at the rate of one and one-half per bump. Two men can mix and apply the asphalt in five 6-hour shifts.

Cost: Modifications can be completed at an estimated $800.00 as shown in the following table:

Speed Bump Modification Cost Estimate

Item	Cost ($)	Time	Quantity	Total ($)
Grinder	55.00/wk	2 wk	—	110.00
Mixer	140.00/wk	1 wk	—	140.00
Roller	10.00/wk	1 wk	—	10.00
Asphalt Mix	2.00/55 lb bag	—	25	50.00
Pebbles	70.00/load	—	1	70.00
			TOTAL	800.00

Advantages:
1. Speed bumps will still deter speeding.
2. Damage to automobiles should decrease.
3. Project can be accomplished by existing staff.
4. Project can be accomplished prior to Term I traffic.

Action:
1. Approve proposal by July 15.
2. Arrange for rental equipment and supplies by August 1.
3. Modify speed bumps August 1–15 while the campus is closed to regular classes.

3. Title the proposal in Exercise 2 and compose a title page. You are the Supervisor of the Maintenance Department of your college presenting the proposal to the campus provost or dean.

4. Write a memo of transmittal. Include a brief abstract of the proposal. Acknowledge an individual or company for assistance in obtaining equipment and supply prices.
5. Devise a Table of Contents.

WRITING OPTIONS

Write a long proposal to solve a specific problem on your campus or at your place of employment. Select a subject of sufficient complexity to warrant a formal report, but do not tackle something beyond your scope. Your proposal should require some research (obtaining information on equipment capabilities, cost estimates, devising a procedure to document the problem). After you complete your proposal draft, prepare all of the appropriate formal report features: title page, transmittal correspondence, table of contents, list of illustrations, abstract, and supplements. Submit your proposal in a folder. Suggested subjects are:

Campus	*Work*
new equipment	new equipment
new system of registration	new uniforms
improved parking facilities	improved system of duty roster
picnic or stone tables	improved method of displaying wares
snack bars by classrooms	index for wage increases
additional computer lab time	improved lounge facilities
additional study carels	improved working conditions
new special interest club	new in-service training program
Saturday or 5:00 P.M. classes	sports team program
a women's center	fire evacuation procedures
a campus jogging path	an advertising program
student entertainment project	new staff orientation program
free movie program	grievance procedures
improved bomb scare procedures	dental insurance program
other?	other?

7

MANUALS

DUFFY

by Bruce Hammond

Skills:

After studying this chapter, you should be able to

1. Recognize the purpose and uses of various types of manuals.
2. Grasp the importance of appropriate language for different manual audiences.
3. Know the procedures for preparing an operations manual.
4. Understand the four major writing tasks of manuals.
5. Critique an operations manual.

INTRODUCTION

Manuals are written guides or reference materials which are used for training, assembling mechanisms, operating machinery or equipment, servicing products, or repairing products. Typically a manual includes

- Precise definitions
- Descriptions of mechanisms
- Step-by-step instructions
- Analyses of processes

In a large company professional or technical writers may prepare manuals for the operation and repair of products or equipment. In smaller companies the preparation, editing, and component-parts review may fall to all competent writers. Those employees who use in-house manuals for training, procedures, and assembling are continually encouraged to review and improve the component parts. This chapter will consider the operation manual because it is the most common. Chapters 8, 9, 10, and 11 will consider the writing tasks of definition, description, instruction, and process analysis.

AUDIENCE

A major consideration for manual writing is audience. The language and technical detail must fit the intended user. Frequently the user of a product manual is a novice or layperson; thus, he or she has a need for definitions, mechanism description, operating instructions, and an analysis of possible problems in a product's operation. Although the product may be a highly complex mechanism, the manual information must be simple, clear, and accurate. Writers of such operation manuals often aim their writing at the comprehension level of an eighth-grade student. Technical terms, abbreviations, symbols, and mathematic procedures are simplified or avoided. If the audience is a group of skilled technicians, the language, of course, may be more technical; however, the writer must be alert to provide definitions, explanations of theory, and graphic reinforcement.

The following two examples illustrate language differences in manuals intended for two different audiences. The first example shows the introductory information from an operation manual for a personal pager. The intended audience is the layman purchaser:

Introduction

Congratulations! You are now using the world's first microprocessor-controlled 900MHz pager. Motorola's advanced technology offers unique features and benefits which provide the ultimate in performance and reliability.

The Dimension 1000 pager is a versatile unit that is designed to provide reliable communications for a variety of applications. To get the full benefit from the pager, please read these operating instructions carefully.

Coding Data Label

Dimension 1000 pagers come in several model configurations equipped with a variety of options which affect the operation of your pager. To determine how your pager operates, refer to the coding-data label located under the belt clip. The pager is capable of one, two, or three calls, depending upon how it was ordered from the factory. The coding-data label indicates the number and type of calls (Figure 2). Tone-only calls are indicated by a "T," and voice calls are indicated by a "V." If a particular call is not present, that area will be blank.[1]

The next example contains the introductory material from an instruction manual for servicing a walkie-talkie radio. The intended audience is skilled service technicians:

Introduction

The MX 300-T "Handi-Talkie" radio described in this manual is the most advanced two-way radio available. Hybrid modular construction is used throughout, reflecting the latest achievements in microelectronic technol-

[1] Motorola, Inc., *Motorola Dimension 1000 Binary GSC Pager* (Fort Lauderdale, Florida: Motorola, Inc., Paging Products Division, 1982), a manual.

ogy. The plug-in modules provide greater flexibility, greater reliability, and easier maintenance.

Each radio contains plug-in hybrid modules. These modules contain over 90% of the electronics—providing faster service and less down-time. Guide pins are provided on the modules to assist replacement and prevent incorrect insertion. Instead of complex wiring harnesses, printed flexible circuits are used in the radio. These durable, thin plastic films eliminate broken, pinched, or frayed wires—with a neat, easy-to service interior.[2]

The first example is written with a "you" perspective. The language is general and nonspecific. The second example is less personal and contains such technical terms as *hybrid modular construction, plug-in hybrid modules, down-time,* and *flexible circuits,* terms familar to technicians.

MANUAL PREPARATION

Let us consider the writer's procedural steps for the preparation of an operations manual for a highly technical product.

Step 1—Determining audience. The writer must determine if the audience consists of laypersons or skilled technicians. The audience will have a bearing not only on the language, but also on the complexity of the graphics, the extent and scope of the data, and the manual size and format. If the intended manual user is a layman, the writer must consult the marketing department to determine who will buy the product, what language is appropriate, and what detail is essential.

Step 2—Consulting the engineering department. Engineers of various specialties must be consulted to determine how the product works, what instructions and warnings must be stressed, and how much detail should be included for different audiences. Considerable more detail, graphics, schematics, and diagrams will be included in manuals for technicians than in those for laypersons.

Step 3—Deciding on manual production. Once the audience and the intent of the manual is established, the writer must make initial decisions on the content and appearance of the manual. He must prepare a rough draft of the content, deciding on headings and numbering systems and on whether or not to use a title page, table of contents, and so forth. Bearing in mind the four common writing components of manuals (definition, description, instructions, and process analysis), the writer must decide on the emphasis and amount of each to include.

[2] Motorola, Inc.. *Motorola MX300-T Five Channel "Handi-Talkie" Portable Radio* (Fort Lauderdale, Florida: Motorola, Inc., Portable Products Division, 1982), a manual.

He must decide if the manual shall be a folded pamphlet which can be inserted into the product packaging or if it will be a booklet with bound pages. He must decide on the paper quality, type of print, color usage, and number of manuals.

Step 4—Preparing the graphics. Although graphic consideration goes hand-in-hand with Step 3, it is important enough to warrant separate discussion. While actual photographs may be used to depict the overall product, drawings are used more extensively in manuals because they can emphasize parts and relationships by the judicious use of exploded, cutaway, and schematic sketches. Lists of items or features should be incorporated into tables. The writer usually devises rough sketches of the desired graphics even though professional graphic artists will prepare the actual art work.

Step 5—Reviewing the copy. Once the copy is prepared, the competent writer reviews the manual carefully with marketing and engineering personnel. Editing the work is essential for a successful manual.

Step 6—Producing the manual. Finally, the copy is given to typesetters who prepare camera-ready copy, print it, and bind or fold the manuals.

Figure 7.1 shows an operation manual for a Motorola Dimension 1000 Binary SC Pager. The original manual was printed as a 2⅝ in. by 3⅝ in., 12-panel, folded booklet designed to fit the packaging of the pager:

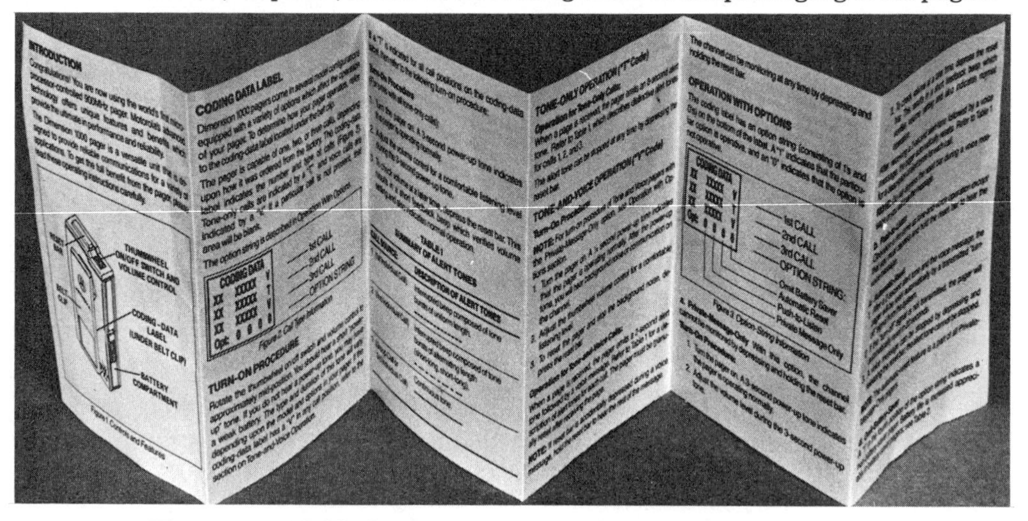

Figure 7.1 Folded operation manual for personal pager (Courtesy of Motorola Inc., Fort Lauderdale, Florida)

To familiarize you with a typical product operation manual, we include the Motorola pager manual in full in Figure 7.2, along with explanatory comments:

Company name & trademark plus name of product

MOTOROLA **DIMENSION 1000**

Binary GSC Pager

Actual photograph provides graphic *description* of product.

MOTOROLA DIMENSION 1000
MICROPROCESSOR CONTROLLED

Title states purpose of manual.

Operating Instructions

68P81025C65-O

Figure 7.2 Operation manual for a Motorola Dimension 1000 Binary GSC pager (Courtesy of Motorola, Inc., Fort Lauderdale, Florida)

The "you" identifies audience.

INTRODUCTION

Congratulations! You are now using the world's first microprocessor-controlled 900MHz pager. Motorola's advanced technology offers unique features and benefits which provide the ultimate in performance and reliability.

Brief mechanism *definition* is followed by manual purpose.

The Dimension 1000 pager is a versatile unit that is designed to provide reliable communications for a variety of applications. To get the full benefit from the pager, please read these operating instructions carefully.

Drawing of overall mechanism labels parts to familiarize audience with controls and features.

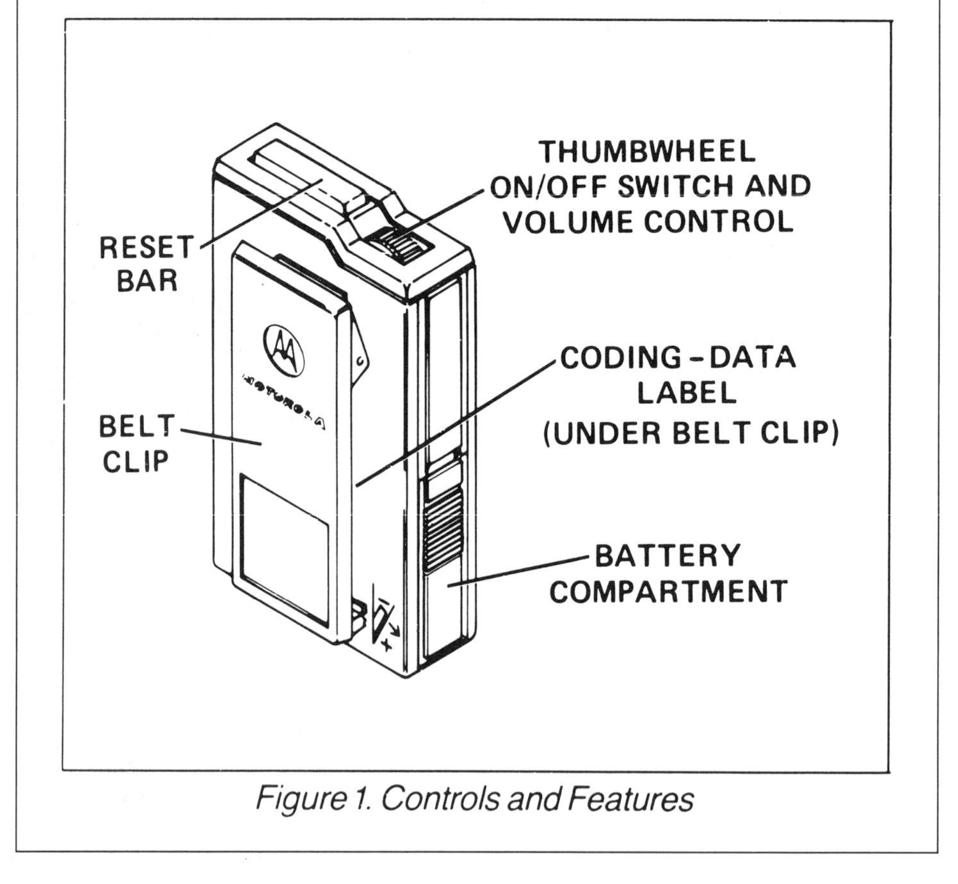

Figure 1. Controls and Features

Figure 7.2 Continued

Text provides brief *analysis* of operation.

CODING DATA LABEL

Dimension 1000 pagers come in several model configurations equipped with a variety of options which affect the operation of your pager. To determine how your pager operates, refer to the coding-data label located under the belt clip.

The pager is capable of one, two, or three calls, depending upon how it was ordered from the factory. The coding-data label indicates the number and type of calls (Figure 2). Tone-only calls are indicated by a "T," and voice calls are indicated by a "V." If a particular call is not present, that area will be blank.

The option string is described in Operation With Options.

Graphic blow-up is more effective than an actual photograph.

CODING DATA			
XX	XXXXX	V	1st CALL
XX	XXXXX	T	2nd CALL
XX	XXXXX	V	3rd CALL
Opt:	0 0 0 0		OPTION STRING

Figure 2. Call Type Information

TURN-ON PROCEDURE

Paragraph provides simple *instructions* along with *analysis* of a problem.

Rotate the thumbwheel on/off switch and volume control to approximately mid-position. You should hear a short "power-up" tone. If you do not hear a power-up tone, you may have a weak battery. The type and duration of this tone will vary, depending upon the model and options in your pager. If the coding-data label has a "V" in any call position, refer to the section on Tone-and-Voice Operation.

Figure 7.2 Continued

Instructions emphasize action (*turn, adjust, depress*).

If a "T" is indicated for all call positions on the coding-data label, then refer to the following turn-on procedure:

Turn-On Procedure

(For units with all tone-only calls):

1. Turn the pager on. A 3-second power-up tone indicates the pager is operating normally.
2. Adjust the volume control for a comfortable listening level during the 3-second power-up tone.
3. To check volume at a later time, depress the reset bar. This results in a short feedback beep which verifies volume setting and also indicates normal operation.

Alert Tones are summarized in a formal table for easy reference.

TABLE 1
SUMMARY OF ALERT TONES

CALL SOURCE:	DESCRIPTION OF ALERT TONES
1 (1st Individual Call)	Interrupted beep composed of tone bursts of uniform length. - - - - - - - - - - - - - - -
2 (2nd Individual Call)	Interrupted beep composed of tone bursts of alternating length (short-long, short-long). - — - — - — - —
3 (Group Call or 3rd Individual Call)	Continuous tone. ————————

Figure 7.2 Continued

Section provides brief *analysis* of operation.

TONE-ONLY OPERATION ("T" Code)

Operation for Tone-Only Calls:

When a page is received, the pager emits an 8-second alert tone. Refer to Table 1, which describes distinctive alert tones for calls 1, 2, and 3.

The alert tone can be stopped at any time by depressing the reset bar.

TONE-AND-VOICE OPERATION ("V" Code)

Turn-On Procedure:

The word *note* is used for emphasis.

NOTE: For turn-on procedure of Tone and Voice pagers with the Private-Message-Only option, see Operation with Options section.

Numerical *instructions* use action verbs (*turn, adjust, depress*).

1. Turn the pager on. A ¼ second power-up tone indicates that the pager is operating normally. After the power-up tone, you will hear background noise or communication on the channel.
2. Adjust the thumbwheel volume control for a comfortable listening level.
3. To reset the pager and stop the background noise, depress the reset bar.

Operation for Tone-and-Voice Calls:

Paragraph provides brief *analysis* of operation.

When a page is received, the pager emits a 2-second alert tone followed by a voice message. Refer to Table 1 for a description of alert tones for each call. The pager must be manually reset after receiving the page.

The word *Note* is used for emphasis.

NOTE: If reset bar is accidentally depressed during a voice message, hold the reset bar to hear the rest of the message.

Figure 7.2 Continued

Graphic blow-up is more effective than an actual photograph.

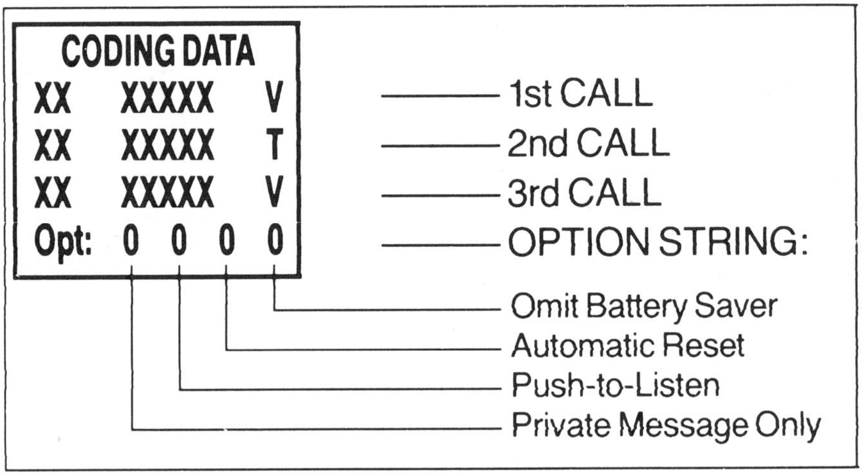

The channel can be monitoring at any time by depressing and holding the reset bar.

OPERATION WITH OPTIONS

The coding label has an option string (consisting of 1's and 0's) on the bottom of the label. A "1" indicates that the particular option is operative, and an "0" indicates that the option is not operative.

CODING DATA

XX	XXXXX	V		1st CALL
XX	XXXXX	T		2nd CALL
XX	XXXXX	V		3rd CALL
Opt:	0 0 0	0		OPTION STRING:

Option string (right to left): Omit Battery Saver; Automatic Reset; Push-to-Listen; Private Message Only

Figure 3. Option-String Information

a. Private-Message-Only With this option, the channel cannot be monitored by depressing and holding the reset bar.

Instructions are continued.

Turn-On Procedure:

1. Turn the pager on. A 3-second power-up tone indicates the pager is operating normally.
2. Adjust the volume level during the 3-second power-up tone.

Figure 7.2 Continued

Instructions are continued with use of *note* to provide emphasis.

3. To check volume at a later time, depress the reset bar. This results in a short feedback beep which verifies volume setting and also indicates normal operation.

After receiving a 2-second alert-tone followed by a voice message, the pager automatically resets. Refer to Table 1 for a description of alert tones for each call.

NOTE: Do not depress the reset bar during a voice message or you will lose the message.

b. Push-to-Listen

Operation is the same as tone-and-voice operation *except* that you must depress and hold the reset bar to hear the voice message.

c. Automatic-Reset

1. After a 2-second alert tone and the voice message, the pager will be reset automatically by a transmitted "turn-off" signal from the system.
2. If the "turn-off" signal is not transmitted, the pager will automatically reset in 20 seconds.
3. A voice message can be stopped by depressing and releasing the reset bar; the alert tone cannot be stopped.

NOTE: The automatic-reset feature is a part of all Private-Message-Only pagers.

Brief analysis is provided.

d. Omit-Battery-Saver

A "1" in the fourth position of the option string indicates a non-battery-saver pager. Battery life is increased appreciably in battery-saver pagers; see Table 2.

Figure 7.2 Continued

e. **Fixed Alert** (not indicated in option string)
The alert tone is always at maximum volume, but the voice-message volume can be adjusted.

Battery information *analysis* precedes instructions for battery changing.

BATTERY INFORMATION

LOW-BATTERY ALERT

Whenever the pager is reset, the battery voltage is monitored. If the battery level is low, a special alternating high-low alert tone will sound. If the low-battery alert sounds, replace the mercury battery or recharge the nickel-cadmium battery. Battery life remaining after a low-battery alert is approximately three hours for nickel-cadmium batteries and eight hours for mercury batteries.

Battery Life: Battery life will vary depending upon the system. The following figures are given as a general indication of expected battery life.

Graphic table provides easy reference data.

TABLE 2
BATTERY LIFE (Approximate)

BATTERY TYPE	BATTERY-SAVER	NON-BATTERY-SAVER
Mercury	200-220 Hours	64-70 Hours
Nickel-Cadmium	66-73 Hours	20-22 Hours

***Instructions* stress action verbs and employ boldface print for emphasis.**

Battery Installation or Replacement:

1. Locate the battery compartment on the side (closest to volume-control thumbwheel) as shown in Figure 4(a).
2. **Slide the ribbed latch down** so that the red indicator is visible.

Figure 7.2 Continued

Numbered instructions continue.

3. Pull open the battery-compartment door (hinged at the bottom) as shown in Figure 4(b). Remove the old battery.
4. Place the positive end of the new battery into the compartment (see polarity marking on the housing shown in Figure 4(b). Push the battery down until it makes contact with the + battery contact.
5. Close the battery-compartment door and slide the ribbed latch up until the red indicator no longer appears.
6. If the battery is an approved rechargeable nickel-cadmium type, charge the battery in the pager as described in the Battery Charging paragraph.

Blow-up drawings assist the written instructions.

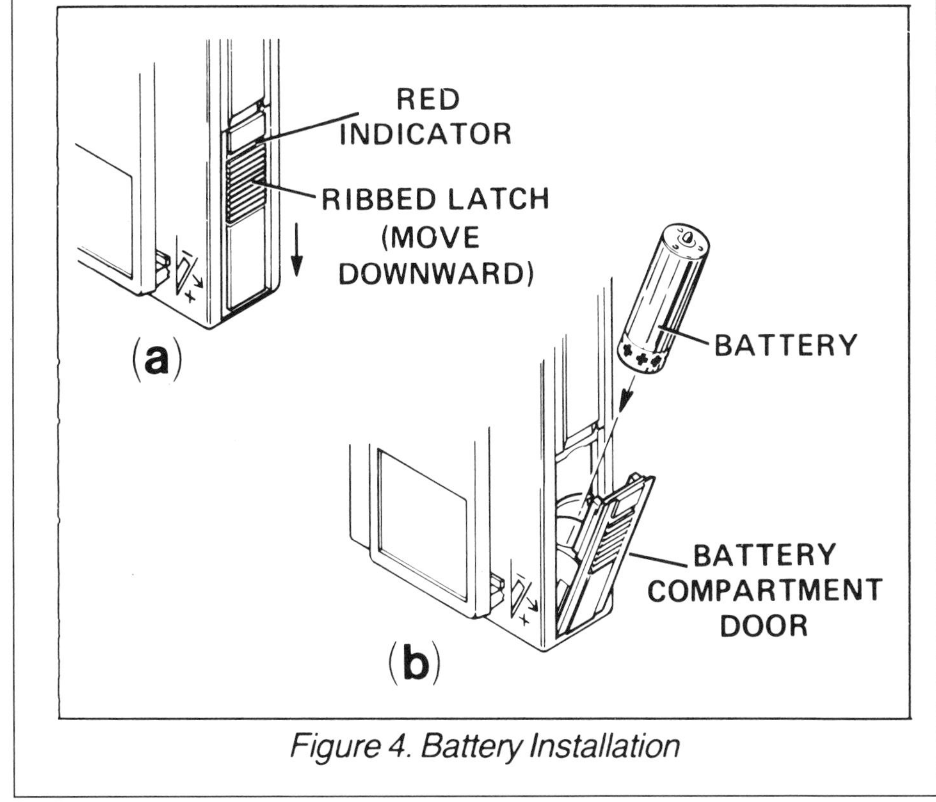

Figure 4. Battery Installation

Figure 7.2 Continued

Explanation actually provides instructions.

Battery Types:

Either disposable mercury or rechargeable nickel-cadmium batteries are available.

Mercury Battery Model: 1.4V N-Size, Motorola NLN6199A

Nickel-Cadmium Model: 1.3V N-Size rechargeable, Motorola Type NLN6965A

Pager performance and battery life cannot be guaranteed if other types of batteries are used.

Brief *analysis* is followed by numbered *instructions*.

Battery Charging:

(Approved Motorola nickel-cadmium battery only)

The Motorola charger (either single or multiple-unit models) charges a pager and a spare N-cell size nickel-cadmium battery simultaneously.

1. Operate the charger from the correct line voltage as indicated on the charger.

2. Insert the pager in the charger with pager power (thumbwheel switch) set to off. For best results, charge the pager at least 12 hours (overnight).

3. A separate nickel-cadmium battery may be charged also by inserting the battery in the auxiliary charging compartment. Observe the correct battery polarity orientation.

The word *note* provides emphasis.

NOTE: The charger is designed to prevent accidental charging of mercury batteries.

Figure 7.2 Continued

Informal table is *instructional.*

BATTERY CHARGERS:

These Motorola single and multiple-unit chargers are available for recharging nickel-cadmium batteries.

NLN4508B	Single-Unit Charger, 117VAC
NLN4509B	Single-Unit Charger, 230VAC
NLN4510B	Multiple-Unit Master Charger (5 receptacles), 117VAC
NLN4511B	Multiple-Unit Auxiliary Charger (6 receptacles); three units can be powered from multiple-unit master (NLN4510B)

Figure 7.2 Continued

Repair and maintenance note provides *instructions.*

REPAIR AND MAINTENANCE

The Dimension 1000 pager, properly handled, will provide years of service. However, should it require repair, Motorola's National Service Organization, staffed with specially trained technicians, offers strategically located repair and maintenance facilities. Consult your Motorola sales representative for service locations in your area.

Copyright information is standard.

COMPUTER SOFTWARE COPYRIGHTS

The Motorola products described in this manual may include copyrighted Motorola computer programs stored in semiconductor memories or other mediums. Laws in the United States and foreign countries preserve for Motorola certain exclusive rights for copyrighted computer programs, including the exclusive right to copy or reproduce in any form the copyrighted computer program. Accordingly, any copyrighted Motorola computer programs contained in the Motorola products described in this manual may not be copied or reproduced in any manner without the express written permission of Motorola, Inc. Furthermore, the purchase of Motorola products shall not be deemed to grant either directly or indirectly or by implication, estoppel, or otherwise, any license under the copyrights, patents or patent applications of Motorola, except for the normal non-exclusive, royalty free license to use that arises by operation of law in the sale of a product.

, Motorola, and Dimension 1000 are trademarks of Motorola, Inc.

Figure 7.2 Continued

EXERCISES

1. Collect three or four operation manuals for such products as a digital clock with alarm, a telephone answerer, a camera, a walkie-talkie, an electric coffee maker, or an electric can opener. Identify in the margins which type of writing is stressed in each section: definition, description of the mechanism, instructions, or analysis.

WRITING OPTION

1. Select the best manual (see exercise) and write a brief report on its effectiveness. Comment on its overall format (size, type of paper, layout, headings, color, boldface print, numbering system), its language, its graphics, the definitions, the descriptions of the mechanism, the instructions, the analysis of operating procedures or problems. Use headings in your own report. Attach the manual or photocopied pages.

8

DEFINING TERMS

Skills:

After studying this chapter, you should be able to

1. Understand the need for definition and explain the methods by which definitions may be incorporated into a report.
2. Write formal sentence definitions for a list of related terms.
3. Recognize and avoid the fallacies in formal definitions.
4. Write an extended definition employing six or more writing strategies.
5. Supplement written definitions with appropriate graphics.

INTRODUCTION

A writer must always be alert to the need for a definition of a term or phrase in all reports. A definition is a statement of the meaning of a word or word group or a sign or a symbol. The need for precise definitions is so crucial that sometimes a separate section of a report or manual is devoted to lists of terms and their meanings.

Every field of endeavor has its own vocabulary. Allied health students must master precise definitions of specific diseases, disorders, procedures, and other medical terminology. Electronics study demands an understanding of *capacitance, resistance, frequency,* and other related terms. In recent years anyone who has shopped for a home computer has been suddenly introduced to a mind-boggling new vocabulary including *byte, floppy disk, computer friendly,* and *CP/M,* to name only a few terms. In fact, the computer age has ushered in many new terms, including *cyberphrenetic*, or one who is excessively excited and fanatical about the study of computers.

AUDIENCE

Audience determines the need for and the extent of definitions. Writers of informal or formal reports must define all terms which may be unfamiliar to the audience, words which may have more than one meaning, or those which are used in a special or stipulative manner.

Consider the simple word *tongue*, which has at least nine distinct meanings to different audiences. To a biologist, a tongue is a fleshy, movable portion of the floor of the mouth of most vertebrates that bears sensory end organs and small glands and which functions in taking and swallowing food, and in man as a speech organ. To a geographer, however, a tongue is a long, narrow strip of land projecting into a body of water. To a cobbler, a tongue is a flap under the lacing or buckle of a shoe at the throat of the vamp. To a linguist, a tongue is the spoken language, manner or quality of utterance, or the intention of a speaker. To a belt maker, a tongue is a movable pin in a buckle. To a carpenter, a tongue is the rib or one edge of a board that fits into a corresponding groove in an edge of another board to make a joint flush. Yet, to a bell manufacturer, a tongue is a metal ball suspended inside a bell so as to strike against the side as the bell is swung. And finally, to some religious groups, tongue is the charismatic (divinely inspired) gift of ecstatic speech. The parenthetical definition of *charismatic* underscores the need for definitions as an essential to meaningful communication.

Definitions may be integrated into a report, may constitute a major portion of a report, or may be the entire report. An instruction manual may include an introductory list of terms to be used within the instructions. A policy handbook may introduce each chapter with relevant definitions. Sometimes an appendix of terms must accompany a report.

METHODS

The extent to which a term should be defined depends not only on the audience but also on the complexity of the term itself. Terms may be defined by

- Parenthetical expression
- Brief phrase
- Formal sentences
- Extended paragraphs

Parenthetical Definition

The simplest way to define a term is to include a synonym in parentheses directly after the term:

> The top half of a drainage map drawing is the plan (aerial view); the bottom half is the profile (horizontal view).
>
> The ring top (round, spoked, carrying handle) of the fire extinguisher corroded.

Definition by Brief Phrase

Sometimes a defining phrase will clarify your term:

> If the body temperature is abnormal, that is above or below a range of 97.6°F to 99°F, further diagnostic procedures should follow.

Formal Sentence Definition

A specific pattern exists for precise definition of terms. The pattern consists of three parts: the name of the term, the class of the term, and the characteristics of the term which distinguish it from all other members of its class. Some examples follow:

Term	*Class*	*Distinguishing characteristics*
Arbitration	is a process	by which both parties to a labor dispute agree to submit the dispute to a third party for binding decision.
Sediment	is matter	which settles to the bottom of a liquid.
Assets	are items owned	such as cash, receivables, inventories, equipment, land, and buildings.
Paranoia	is a personality disorder	in which a person feels persecuted or has ambitions of grandeur.
Arson	is a criminal act	of purposely setting fires to a building or property so as to collect insurance.
A patent	is an inventor's exclusive right	which is granted by the federal government, to own, use, make, sell, or dispose of an invention for a certain number of years.
Down-time	is a period	during which a computer system is inoperable due to power failure or hardware breakdown.

Sometimes a formal definition requires more than one sentence in order to read smoothly:

> A rifle is a firearm which has spiral grooves inside its barrel to impart a rotary motion to its projectile. It is designed to be fired from the shoulder and requires two hands for accurate operation.

These definitions concentrate on just one meaning of the term. The term *rifle* is also a verb, *to rifle*, which is "to cut grooves, or to ransack, rob, or pillage, or to search and rob."

To be useful a definition should not contain terms which are more technical or confusing than the term itself. Consider which of these two definitions is more useful to a layperson:

> Dysgraphia is a transduction disorder which results from visual motor integration disturbance.
>
> Dysgraphia is a writing disorder which results from a difficulty in writing what one sees.

The first definition might well convey meaning to a group of learning disability specialists, but the second is more helpful to an undergraduate student.

The definition writer must avoid certain definition fallacies (misleading errors). For instance, when naming the term's class, one must be careful not to be too broad. To write "A rifle (term) is an object (class)" is too broad. "A rifle (term) is a firearm (class)" is more precise. Conversely, to classify a term too narrowly as in "A chair is a four-legged seat . . . " restricts the possibility that a chair may have a single pedestal for a base, or have five or more legs arranged in a circle.

It is also important to avoid using any form of the term in the second and third parts of the definition; to do so takes your reader in a circle. Consider this statement: "A radical is a person having radical views concerning social order and systems." Does it define *radical*? Other examples of circular definition are

> Comatose is the state of being in a coma.
>
> Fertilization is the process of fertilizing.
>
> A surveyor is one who surveys.

Finally, one must avoid the terms *when* and *where* in formal definition. To write "A crypt is where one is buried" fails to classify the term as a subterranean chamber or to distinguish it from a mausoleum or a cemetery, other places where people are buried. To write "Osmosis is when fluid passes through a membrane" fails to point out that osmosis is a digestive process.

In summary, when devising formal definitions *do not* (1) use needlessly technical language, (2) employ classifications which are too broad, (3) employ classifications or list characteristics which are too narrow, (4) use any form of the term itself in the class or characteristics citation, or (5) use the terms *where* or *when* in place of a classification.

EXTENDED DEFINITION

If your intent is to define a term so that your audience has a thorough understanding of it, you may need to extend your formal definition. You may want to include physical description, examples, synonyms, comparisons or contrasts, analogies, explanations of the origin of the item, an etymology (origin) of the term itself, discussions of cause and effect, or analysis of the involved process. Such a definition is called an *extended definition*.

The extended definition may be a few sentences, a paragraph, several pages, or an entire lengthy report.

Should you decide that an extended definition is necessary, a number of strategies are at hand to add to your formal definition. Select as many of the methods as are necessary to clarify completely the meaning of your term. There is no formula for the number or order of the methods to be employed.

Description

Following your formal definition, which distinguishes the item from other members of its class, it may be wise to describe the physical parts of the item, if, indeed, the term names an object. The following example begins with a formal definition and includes a brief physical description:

> An otoscope is a hand-held, diagnostic instrument which is used for examining the external canal of the ear and the ear drum. Composed of plastic, aluminum, and glass, it consists of three main parts: a dry-cell battery barrel assembly, a lens assembly, and a speculum (reflector).

To further describe its parts, subparts, dimensions, weight, and method of use would entail other specific writing strategies which are covered in Chapter 9. A graphic illustration incorporated into the text would help to clarify the term. Figure 8.1 shows the main parts of an otoscope:

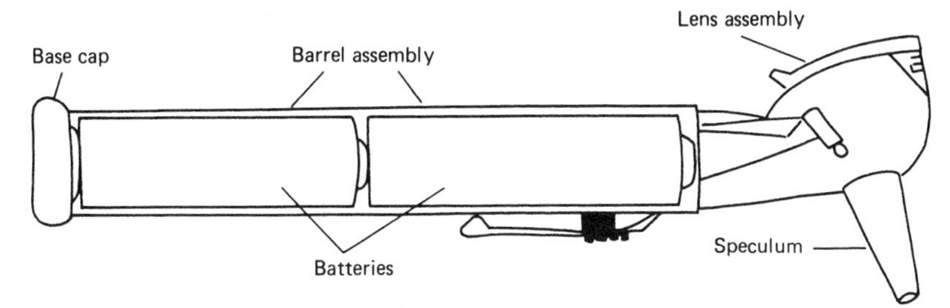

Figure 8.1 The main parts of an otoscope

Examples

A second approach to extended definitions is to provide examples. These may be brief or longer themselves. Two samples follow:

> A parasite is a plant or animal which lives on or within another organism, from which it derives sustenance or protection without making compensation. Some parasites are tapeworms, sheep ticks, lice, and scabies.
>
> A market is a state of trade which is determined by prices, supply, and demand. In salesmanship the term *market* may refer to a trade or commerce in a specific service or commodity, such as the housing market, the stock market, or the designer jean market. Investors in a market study it carefully before investing. In the stock market, a potential investor studies the stock market exchange, current rates of stocks, and current prices of stocks to determine if the "market" is going up or down and to determine whether an investment would be profitable at a particular time.

Synonym

A third expansion of the basic definition is to list synonyms for the term. When readers are familiar with the synonym, they can better understand the definition. Here are some examples of synonym extension:

> A law is a rule of conduct which is established and enforced by the authority, legislation, or custom of a given community or other group. *Rule, regulation, precept, statute,* and *ordinance* are all synonyms for the word *law.*
>
> A motive is an inner impulse or reason that causes a person to do something or to act in a certain way. The terms *intent, incentive,* and *inducement* are sometimes used synonymously with *motive.*

One must be careful to remember that no two words are ever exactly synonymous. Meanings may overlap, but a careful examination of the differences is what conveys precise meaning.

Contrast/Negation

A fourth writing strategy for extended definitions is to contrast or to negate the term from those terms or items in the same class with which it may be confused. For example:

> In criminal investigation *motive* and *intent* are not truly synonymous. A man who provides a lethal drug to a terminally ill patient has the "motive " to alleviate suffering. Still his "intent" is to kill, making him criminally liable for the death.
>
> A stalactite is an elongate deposit of carbonate of lime which hangs from the roof or sides of a cave. It is not a stalagmite, which is a cone-shaped deposit of carbonate of lime which extends vertically from the floor of a cave.

We often understand better what a thing is by knowing what it is not.

Comparison

Conversely, we can better understand what some terms denote if we can examine similarities to things more familiar. Two comparisons for extending definitions follow:

> A stalactite resembles an icicle in that both are hanging, cone-shaped deposits formed by dripping water. In the case of the stalactite the water evaporates leaving the lime deposit; in the case of an icicle the dripping water freezes.
>
> Resistance, the opposition to the passage of electrical current which converts electric energy into heat, may be compared to friction between any two objects. When a sulfur-coated matchstick is rubbed against an abrasive surface, the friction creates a spark which ignites the sulfur and produces heat.

Analogy

The extended comparison of two otherwise dissimilar things is an analogy. Although a computer and a weather thermometer are basically different, the following analogy helps us to understand an analog computer:

> An analog computer is an electronic machine which translates measurements, such as temperature, pressure, angular position, or voltage into related mechanical or electrical quantities. The operating principle of an analog computer may be compared to that of an ordinary weather thermometer. As the weather becomes cooler or warmer, the mercury in the glass tube rises or falls. The expansion and contraction of the mercury has a relationship to the conditions of the weather. The thermometer provides a continuous measurement that corresponds to the climatic temperature. The analog computer makes continuous scientific computations, solves equations, and controls manufacturing processes.[1]

[1]Raymond E. Glos and others, *Business: Its Nature and Environment*, 9th ed. (Cincinnati, Ohio: South-Western Publishing Company, 1980), p. 457.

Origin

Examining the source of an item helps us to grasp the meaning of the word. The following passage briefly explains the source, the mining, and the metallurgy of tin:

> The earliest known tin is found in bronze (a copper-tin alloy) items found at Ur, dated about 3500 B.C. Tin is an element which occurs in cassiterite deposits. The ore is recovered by both opencut and underground mining. The smelting processes include roasting and leaching in acid to remove all of the impurities. The crude tin is resmelted and then refined by further heat treatments of two steps: liquation or sweating, and boiling or poling. Finally, the pure tin is cast in the form of 100-lb ingots in cast iron molds.

Etymology

Examining the origin of a term, its etymology, also helps to convey meaning. Both standard and specialized dictionaries provide etymologies of almost all words in modern English usage; the following are some:

> The term *cyclone* is derived from the Greek word *kykloma* which means "wheel" or "coil."
>
> The term *gerrymander,* the practice of dividing a voting area in such a way as to give unfair advantage to one political party, is derived from Elbridge *Gerry,* the governor of Massachusetts in 1812 when the method was employed, plus the word *salamander* which describes the shape of the redistricted Essex County.

Cause/Effect

To understand completely some terms, an examination of causes and effects is useful. The following example describes the causes and effects of a tornado to amplify the formal sentence definition:

> A tornado is a storm characterized by a violently rotating funnel cloud which has a narrow bottom tending to reach to the earth. The cloud may rotate clockwise or counterclockwise at approximately 100–150 mph. The funnel cloud results from the condensation of moisture through cooling by expansion and lifting of air in the vortex. The air outside of the funnel cloud is also part of the vortex, and near the ground this outer ring becomes visibly laden with dust and debris. Although a tornado takes only a minute or so to pass, it results in devastating destruction. Buildings may be entirely flattened, exploded to bits, or moved for hundreds of yards. Straws are known to be driven through posts. The roar of a tornado can be heard as far as 25 miles away.

Process Analysis

A final strategy of expanded definition is to analyze the process in which the item is involved:

> A skeleton is the bony framework of any vertebrate animal. It gives the body shape, protects soft tissue and organs, and provides a system of levers, operated by muscle, that enables the body to move. Bones of a skeleton store inorganic sodium, calcium, and phosphorous and release them into the blood. The skeleton houses bone marrow, the blood-forming tissue. Bones are joined to adjacent bones by joints. The bones fit together and are held in place by bands of flexible tissue called *ligaments*.
>
> A pressure cooker is an airtight metal container which is used for quick food preparation by means of steam under pressure. When the lid of the pot is secured by means of a rubber gasket and the container is placed on a heat source, fast-flying molecules of steam constantly bump against each other and the inside surface of the container. The combined blows from all molecules exert heat and pressure, "cooking" the food in approximately one-third of the times required by conventional cooking methods.

Process analysis may be extended into a separate writing strategy and is discussed thoroughly in Chapter 11.

Figures 8.2 and 8.3 contain two sample extended definitions. Each begins with a formal sentence definition and utilizes several extended-definition writing strategies. Brief graphics enhance the definitions. The writing strategies are noted in the left margin. Both samples are intended for a lay audience. More detailed and complex definitions would be necessary for a law textbook on patents or for a medical textbook on cancers. Be prepared to critique the samples in class.

A BUSINESS LAW TERM - PATENT

Formal Definition A patent is a document which grants the monopoly right to an inventor to produce, use, sell, or gain profit from an invention or process for a certain number of years. Upon application by the inventor to the United States **Process** Patent Office, an examination is conducted to determine if the invention is new and patentable under the law. If the invention is original and undisputed, a patent is granted and all documents are open to the public. The **Etymology** term patent is derived from the Latin word partere, meaning "to be open." **Description** In order to be patentable the invention must be new and useful to the field of art or technology to which it is related. Only things, such as machines, **Examples** appliances, fixtures, mechanisms, and chemical compounds can be patented. **Negation** Scientific principles, ideas, ways of doing business, and applications of fissionable material or atomic energy for atomic weapons cannot be patented. **Comparison and Contrast** Patents are similar to copyrights but should not be confused with them. While a copyright protects an author, artist, or publisher against duplication of literary, musical, dramatic, or artistic work and lasts for fifty years after the death of the creator, a patent protects an inventor against the copying of the invention and lasts for only seventeen years without the right to renew. Many products exhibit both patent protection for the

Figure 8.2 Sample extended definition (Courtesy of student Deborah Ekback)

invention and trademark protection for the name. An example of such is Pampers. The patent notice includes the U.S. patent numbers as shown in the following:

*PAMPERS is a registered trademark for a unitized, pleated diaper.

U.S. Patents 3,848,594, and 4,041,951

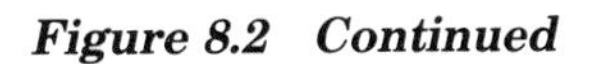

Figure 8.2 Continued

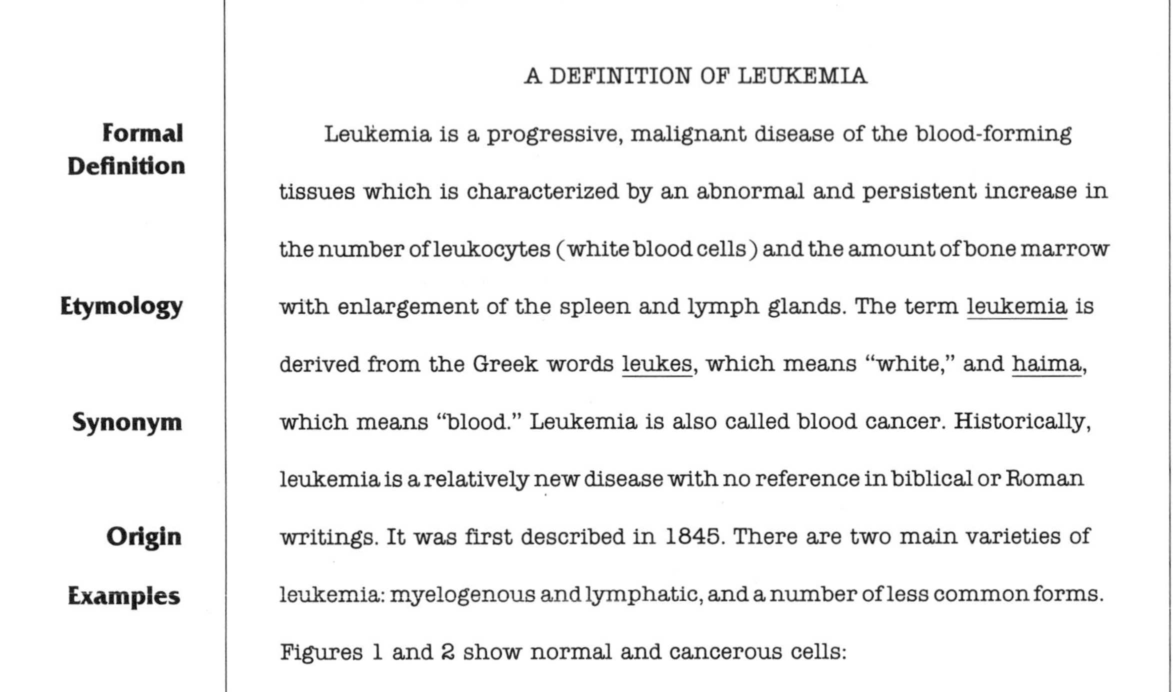

A DEFINITION OF LEUKEMIA

Formal Definition

Leukemia is a progressive, malignant disease of the blood-forming tissues which is characterized by an abnormal and persistent increase in the number of leukocytes (white blood cells) and the amount of bone marrow with enlargement of the spleen and lymph glands.

Etymology

The term leukemia is derived from the Greek words leukes, which means "white," and haima, which means "blood."

Synonym

Leukemia is also called blood cancer.

Origin

Historically, leukemia is a relatively new disease with no reference in biblical or Roman writings. It was first described in 1845.

Examples

There are two main varieties of leukemia: myelogenous and lymphatic, and a number of less common forms. Figures 1 and 2 show normal and cancerous cells:

Descriptive Graphics

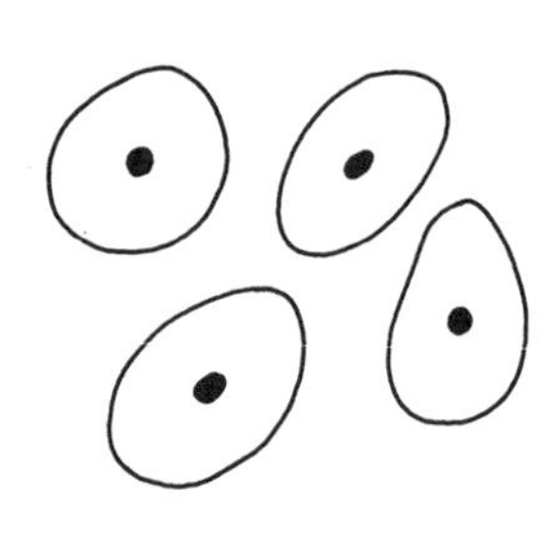

Figure 1 Normal cells

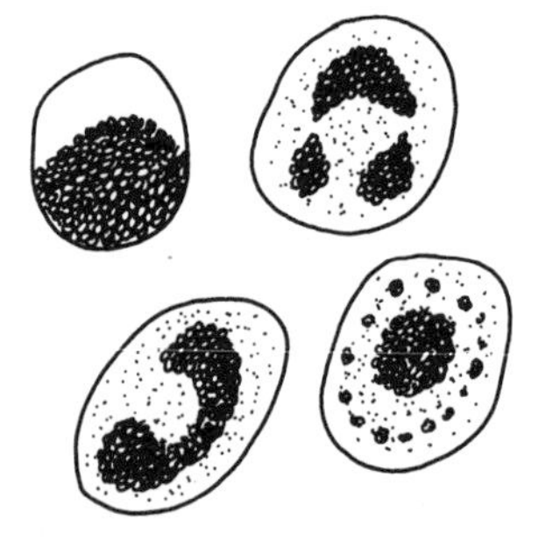

Figure 2 Cancerous cells

Causes

The cause of leukemia is not known although it is known that exposure to radiation is a clear factor. Other causes may include genetic, chemical, viral,

Figure 8.3 Sample extended definition

Effects

and hormonal factors. Usually the first symptoms of leukemia are weakness, increased fatigue, and anemia or hemorrhages. In the acute form symptoms may be severe, the progress rapid, and fever present. In the chronic form early symptoms may be overlooked until enlargement of the spleen or lymph glands is detected. The white blood cell count may increase to 100 times the normal amount, or it may be lower than normal. Ill-formed leukocytes appear in the blood. Although the disease may be treated or the patient can experience spontaneous remission, leukemia is usually fatal.

Figure 8.3 Continued

EXERCISES

1. What is wrong with these formal definitions? Rewrite each by providing a precise class and distinguishing characteristics. Consult dictionaries and specialized encyclopedias.
   - **a.** A latent image is a prephotographic image on a film which cannot be seen.
   - **b.** Cramming is when a student attempts to learn most of the contents of a course in a day or two.
   - **c.** Asthma is a condition of continuous or paroxysmal labored breathing which is accompanied by wheezing, a sense of constriction in the chest, and often attacks of coughing or gasping.
   - **d.** Celluloid is a substance which is thin and inflammable, used for photographic films.
   - **e.** A bond is when two things, such as concrete and steel, are adhered.
   - **f.** A potentiometer is a device used as a resistor.
   - **g.** An azimuth is a 45° angle measured clockwise from any reference meridian.
   - **h.** A journal is a diary in which the analysis of each of the daily financial transactions of a business are recorded as they occur.
   - **i.** A binary cell is a cell in a computer's memory which is restricted to storing binary numbers.
   - **j.** Anxiety is when you are paralyzed with fright and do not know what you are afraid of.
2. In the following extended definition identify each writing strategy by writing the methods in the space provided after each sentence.

   A cyclone is a storm that may range from 50 to 900 miles in diameter and that is characterized by winds of 90 to 130 mph blowing in a circle—counterclockwise in the northern hemisphere and clockwise in the southern hemisphere—around a calm center of low atmospheric pressure while the storm itself moves from 20 to 30 miles per hour. ______________________________ Cyclones may be called whirlwinds, hurricanes, and typhoons. ______________________ The term *hurricane*, however, is properly applied only to a cyclone of large extent and suggests the presence of rain, thunder, and lightning. The term *typhoon* refers to tropical cyclones in the region of the Phillipine Islands or the China Sea. ______________________ A tornado is not a cyclone. Although a tornado consists of whirling winds, it is characterized by a funnel-shaped cloud which is far smaller in diameter than a cyclone and by winds far exceeding the velocity of winds in a cyclone. __________ The term *cyclone* is derived from the Greek word *kykloma* which means "wheel" or "coil." ______________________

WRITING OPTIONS

Wednesday

1. Select five related terms from your professional field. For this report try to avoid terms which name mechanisms. Title your assignment by stipulating the field of the terms: e.g., "Terms Used in Radiation Technology," "Terms Used in Electronics." Develop formal sentence definitions for each of the terms. Each definition must state the term, classify the term, and cite the distinguishing characteristics which differentiate your term from all others in its class. Avoid the fallacies. Following are some suggested terms in specialized fields:

Word Processing	*Computers*
menu	byte
wordwrap	hashing
block move	live screen
global search	CP/M
dedicated function	program-file compability

Fashion	*Electronics*
godet	electron
grommet	resistance
stonewashed	frequency
peplum	capacitance
double-faced linen	Ohm's law

Criminal Justice	*Allied Health*
larceny	emphysema
felony	atherosclerosis
manslaughter	angina
assault	escemia
battery	vasodilation

Fire Science	*Psychology*
arson	anxiety
pyromaniac	psychosis
flammable	neurosis
purple K	schizophrenia
cartridge	manic depressive syndrome

Marketing	***General Business***
a good	sole proprietorship
convenience good	partnership
shopping good	limited partnership
specialty good	corporation
unsought good	conglomerate

Surveying	***Architecture***
azimuth	fascia
stadia	cantilever
transverse	soffit
hub	strut
transit	beam

Political Science	***Astronomy***
democracy	nova
communism	black hole
socialism	albedo
oligarchy	transit
monarchy	solar eclipse

Incorporate graphics into this definition report wherever they are appropriate.

2. Select a broad term (one naming a field of study or a concept) from your professional field. Write an expanded definition of the term which is suitable for first-year students in the field. Begin with a formal sentence definition and then expand it by employing at least five extended definition writing strategies. Employ graphics if possible. Use parenthetical or phrase definitions for unusual terms within your extended definition. In the margin jot down the writing strategies which you have employed. Approximately 250 words.

9

DESCRIBING MECHANISMS

Skills:

After studying this chapter, you should be able to

1. Define *mechanism.*
2. Understand the need for descriptions of mechanisms.
3. Explain the difference between a general and a specific description.
4. Distinguish between a description of a mechanism and instructions for a mechanism's operation.
5. Write a brief, *general* description of a mechanism embodying logical organization, an objective tone, precise terminology, and graphics.
6. Write a detailed, *specific* description of a mechanism embodying logical organization, an objective tone, precise terminology, and graphics.

INTRODUCTION

Written descriptions of the tools, appliances, apparatus, and machines which we purchase or operate are necesary for us to understand thoroughly their function. We may call any object which has functional parts a mechanism. We are considering, then, not only a simple pocket knife or rachet wrench, but also a tape recorder, a camera, a stethoscope, a lawn mower, an automobile, and even body organs.

Writers must describe mechanisms to spur sales, to explain assembly, to instruct on operating and procedures, to explain functions, or to analyze strengths and weaknesses. Such descriptions may appear in textbooks, owner and service manuals (see Chapter 8), merchandise catalogues, specialized encyclopedias, specification catalogues, and do-it-yourself trade books.

Except for sales promotion materials which may involve some subjective writing, descriptions of mechanisms are characterized by objectivity, specificity, and thoroughness. A well-written description should enable a reader to understand the mechanism and the function of its parts. Further, the description should enable the reader to judge the efficiency, reliability, and practicality of the mechanism.

AUDIENCE

The purpose of the description and the audience for whom it is written will dictate the length and the amount of technical detail to be included. A *general* description, written for an encyclopedia or a general how-things-work book will emphasize the overall appearance of the mechanism and its parts and explain its purpose, function, and operation. A *specific* description written for an owner's manual or service manual will emphasize not only an overall description of the mechansism and its parts but also will include a detailed description of each part, subpart, or assembly of parts. In addition, a description of a specific mechanism will discuss its strengths, limitations, optional equipment, and/or similar models to allow the reader to judge the usefulness of the particular brand or model.

ORGANIZATION

Whether your description is general or specific, logical organization will aid your reader. An outline should be developed and followed carefully. There are three major sections to a general description of a mechanism:

1.0 General description, or the mechanism as a whole
2.0 Functional description, or the main parts
3.0 Concluding discussion, or assessment

An outline could be much more detailed. The components of a specific description of a mechanism could be outlined, for example, in the following manner:

1.0 The Mechanism as a whole (introduction)
- **1.1** Intended audience
- **1.2** Formal definition and/or statement of purpose or function
- **1.3** Overall description (with graphics)
- **1.4** Theory (if applicable)
- **1.5** Operation (if applicable, with procedural graphics)
  - **1.5.1** Who
  - **1.5.2** When
  - **1.5.3** Where
  - **1.5.4** How
- **1.6** List of main parts (with labeled graphic)

2.0 The Main Parts (body)
  2.1 Description of first part
    2.1.1 Definition and/or purpose statement
    2.1.2 List of subparts (if an assembly)
    2.1.3 Shape, dimensions, weight (with graphic)
    2.1.4 Material and finish
    2.1.5 Relationship to other parts and method of attachment
  2.2 Description of second part . . . (etc.)
3.0 Concluding discussion/assessment (closing)
  3.1 Advantages
  3.2 Disadvantages
  3.3 Optional uses and equipment
  3.4 Other models (with graphics)
  3.5 Cost
  3.6 Availability

This outline is only a guide. The purpose of your description and the intended audience will suggest the amount of detail needed for each report.

PREFATORY MATERIAL

Titles

A brief, clear, limiting title is the first writing strategy. A specific description will include the brand and model designation in the title.

Examples

Description of a Turbine Bypass Valve
Description of a Japanese K-D Socket Wrench
Description of a Universal Pressure Cooker, Model 4S

Intended Audience

An introductory statement of the intended audience and the purpose of the description may be included.

Examples

This description of a turbine bypass valve is intended for engineering students interested in the general construction, operation, and function of such valves.

This description of an Ace bit brace is intended for a junior high shop class instructional manual.

This description of a K & E pencil-lead holder is intended for a descriptive catalogue of architectural, designer, and drafting supplies.

Definition/Purpose

As the outline indicates, it is logical to include a formal definition and/or statement of purpose and function of the mechanism.

Examples

The pressure cooker is an airtight, metal container which is used to cook food by steam pressure at temperatures up to 250°F.

The K-D socket wrench is a hand tool designed to hold and turn fasteners, such as bolts, nuts, headed screws, and pipe lugs.

It is often helpful to compare the mechanism to something similar which is likely to be more familiar to the reader.

Examples

The pressure cooker resembles an ordinary "dutch oven" pot or a large, covered saucepan.

The heart is like a pump in that both draw in liquid and then cause it to be forced away.

Overall Description

Next the physical characteristics or the mechanism are examined. Include a description of the mechanism's shape and/or dimensions, the weight, the materials from which it is constructed, the color, and the finish. Graphic illustration of the mechanism will help the reader to visualize the mechanism.

Example

The K-D socket wrench is made of variable grades of steel. The handle is etched to provide a firm grip. The wrench shaft is 6½ inches long, and the head is 2 inches deep. It weighs 13 ounces. Figure 1 shows the K-D wrench and its overall dimensions:

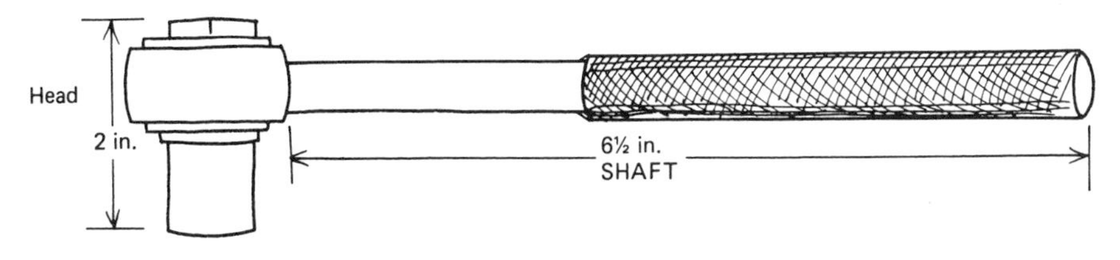

Figure 1 The Japanese K-D Socket Wrench

Theory

If knowledge of theory is essential, it should be included in the introductory or prefatory material.

Examples

The functioning principle of an ordinary mercury theromometer is based on the property of thermal expansion possessed by many substances; that is, they expand when heated and contract when cooled.

The microwave oven cooks food by producing heat directly in the food. As microwaves enter the food, they cause the moisture or liquid in the food to vibrate, and the resulting friction causes the food to heat.

Operator/Process

If the mechanism requires an operator, the qualification or specialty of the operator should be named. Next, the process of the mechanism in action should be explained. An explanation of process should not be confused with instructions which give commands. In explaining the mechanism's process, use third-person subjects and present-tense verbs, in either the active or passive voice.

Examples

Cooks or chefs who wish to extract fresh garlic juice without pulp and skin use the garlic press. The cook (*third person*) places (*present tense, active voice*) the bulb of garlic inside the hollow wedge section of the strainer next to the plate of the press. He (*third person*) squeezes (*present tense, active voice*) the handles together, flattening the garlic and forcing the juice through the small holes of the strainer.

The stethoscope is designed to be used by doctors, nurses, and trained paraprofessionals to convey sounds in the chest or other parts of the body to the ear of the examiner. The earpieces (*third person*) are placed (*present tense, passive voice*) in the examiner's ears. The bell (*third person*) is held (*present tense, passive voice*) against the area of the body to be examined. The sounds (*third person*) are amplified (*present tense, passive voice*) through the tubing by the diaphragm assembly.

A graphic drawing may help the reader to visualize the mechanism in action.

List of Parts

Finally, the main parts of the mechanism are listed. The sequence should have organizational logic. You may list the parts spatially (from outside to inside or top to bottom), functionally (the order in which the parts work), or chronologically (the order in which the parts are put together).

If a part is complex—that is, it contains a number of subparts, such as nut, bolts, springs, pins, and so forth—the part may be called an assembly. Use the following sentence pattern:

Sentence Pattern	The __________ consists of __________ main parts: the __________, the __________, the __________, and the __________.
Spatial Example	The K & E lead holder consists of five main parts: the casing, the push knob, the spring, the tube, and the jaws.
Functional Example	The camera consists of six main parts: the housing assembly, the film feed assembly, the lens, the shutter, the distance setting, and the viewfinder.
Chronological Example	The Schwinn 3-speed bicycle consists of six main parts: the frame and rear wheel assembly, the front wheel assembly, the handlebar assembly, the gear-cable assembly, the brake assembly, and the seat assembly.

THE MAIN PARTS

This section, possibly the lengthiest part of your report, should define and describe each part or assembly in detail. However, a general description of a mechanism will not require as much detail as will a specific description.

Definition/Purpose

Each part requires a formal definition or statement of its purpose. Use one of the following sentence patterns to introduce each part or assembly:

Sentence Patterns	First, the __________ is designed to __________. The __________, the first main part, supports __________. The first functional part, the __________, connects __________.
Examples	First, the etched handle is designed to provide a firm grip. The base, the first main part, supports all of the other parts. The first functional part, the pedestal, connects the base to the hole punch.

If a main part is an assembly, its subparts should be named.

Sentence Pattern	The ______________ assembly consists of the following subparts: the ______________, the ______________, and the ______________.
Example	The direction assembly consists of the following subparts: the tension spring, the pin, and the knob.

Description

Next, the shape, dimensions, and weight of the part are described. If the material and finish of a part differ from the overall description, each should be described. The strategy should be to explain how each part is related to the other parts and how each is attached to the overall mechanism. If the part is an assembly, each subpart should be described in the order listed.

A graphic illustration of each part or assembly may be appropriate. Such graphics may include exploded drawings, sections, or schematics.

Example

The container, the first main part, is designed to hold and to measure the food to be chopped. It is a round, glass bowl which is 4½ inches high and 3½ inches in diameter. The container is etched in 2-ounce gradients, and it has a capacity of 12 ounces (1½ cups). The top rim is threaded to receive the lid. (Figure 2)

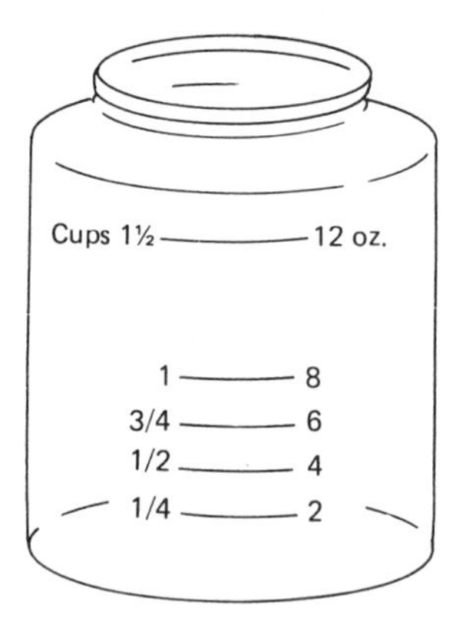

Figure 2 Food chopper container

The plunger assembly, the second main part, consists of the following subparts: shaft, knob cap, shaft housing, shaft housing cap, spring, and blades. When the plunger assembly is depressed, the blades rotate and chop the food in the container.

The shaft is a solid piece of pot metal 8½ inches long and 3⁄16 inches in diameter. It is slightly spatulate at the end where the blades are welded to it. Two stopper tabs protrude 2½ inches from the blade end to secure the shaft in position.

A bulb-shaped, wooden knob cap is pressed securely to the top end. The cap is ⅜ inches long and ½ inch wide.

The shaft housing is a hollow tube 3⅛ inches long and ⅙ inch in diameter. The housing fits over the shaft and contains the spring. A threaded shaft housing cap secures the spring into position.

The steel spring coils around the shaft inside of the housing. The spring is 3 inches long.

The two blades, the final subparts of the plunger assembly, are razor-sharp steel. Each is 2½ inches long and ½ inch high. They are bent at a 45° angle and welded to the spatulate end of the shaft.

Figure 3 shows the plunger assembly parts:

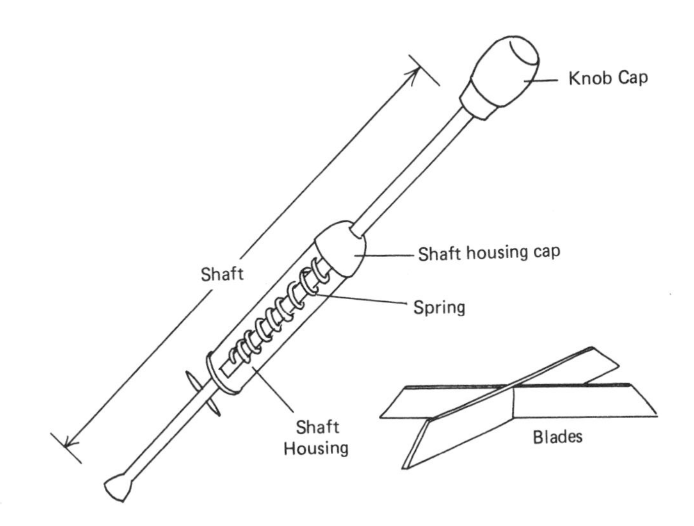

Figure 3 Plunger assembly

A complete description of a manual food chopper would include description of the other main parts: the lid and the chopping pad.

CONCLUDING DISCUSSION/ASSESSMENT

The concluding discussion assesses the efficiency, reliability, and practicality of the mechanism. This assessment may include an examination of the mechanism's advantages and disadvantages, its limitations, its optional uses, the comparison of one model to another model, and the cost and availability.

Example

The Bostich B8 desk stapler is compact and lightweight, making it easy to store and to transport. The finish is scratch resistant and rust proof. It can be used as a tacker as well as a paper stapler.

No more than 20 pages of copy can be stapled at one time. The Bostich Standard stapler is recommended for larger volume.

The recommended retail price is $8.95. A box of 5000 staples is approximately $3.00. The Bostich Standard stapler is sold for $16.75. The Bostich staplers are available in most office supply stores.

Figures 9.1 and 9.2 show sample mechanism descriptions. Figure 9.1 is a general description while Figure 9.2 is a specific mechanism description employing far more detail. Be prepared to critique the samples in class.

A STOP VALVE

Definition and Purpose

A screw-down stop valve is a tap which is used to control the flow of liquids and gases. A water faucet is a stop valve.

Operation of parts

A stem or screw spindle is surmounted by a handwheel. The water flows through an opening whose edge forms the valve seat. The stem has a disc which is usually provided with a replaceable sealing washer to make the actual contact with the seat and thus stop the flow of water.

Parts named in text

To open the tap, the disc is raised by rotating the handwheel in a counter-clockwise direction so that the stem is screwed out of the valve body. Clockwise rotation brings the valve disc into contact with the seat and thus closes the tap. Figures 1 and 2 show the screw-down stop valve in the closed and open positions.

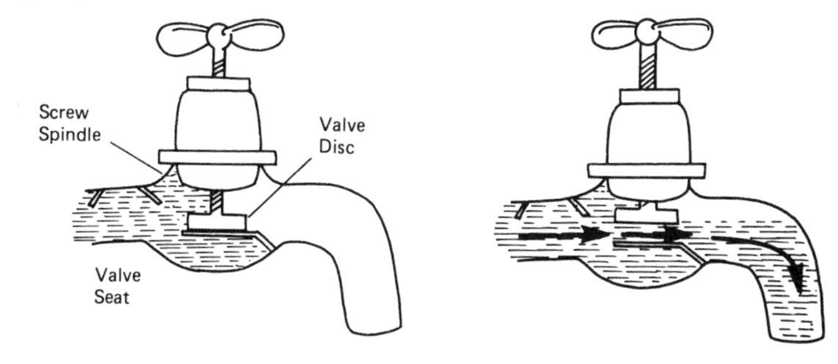

Figure 1 Closed   *Figure 2 Open*

Concluding Discussion

Other types of taps are plug cocks and sluice valves. The screw-down stop valve allows for more accurate control of the rate of flow of the fluid

Figure 9.1 General description of a mechanism

than does the plug cock. Sluice valves, which have stuffing boxes to prevent leakage, are used to control the flow in water mains, pipelines, and so forth. Screw-down stop valves are readily available from plumbing supply houses.

Figure 9.1 Continued

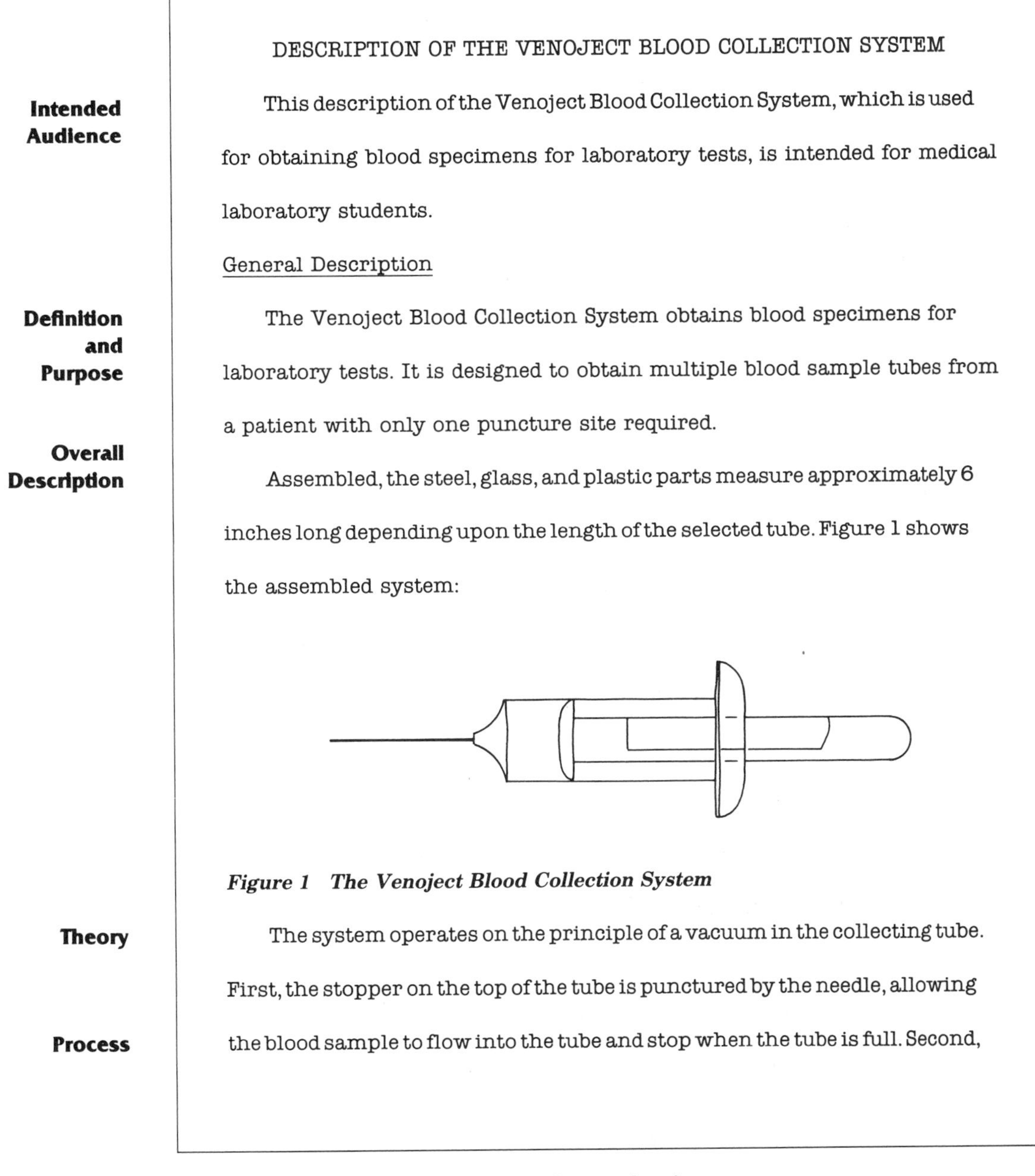

DESCRIPTION OF THE VENOJECT BLOOD COLLECTION SYSTEM

Intended Audience

This description of the Venoject Blood Collection System, which is used for obtaining blood specimens for laboratory tests, is intended for medical laboratory students.

General Description

Definition and Purpose

The Venoject Blood Collection System obtains blood specimens for laboratory tests. It is designed to obtain multiple blood sample tubes from a patient with only one puncture site required.

Overall Description

Assembled, the steel, glass, and plastic parts measure approximately 6 inches long depending upon the length of the selected tube. Figure 1 shows the assembled system:

Figure 1 The Venoject Blood Collection System

Theory

The system operates on the principle of a vacuum in the collecting tube.

Process

First, the stopper on the top of the tube is punctured by the needle, allowing the blood sample to flow into the tube and stop when the tube is full. Second,

Figure 9.2 Specific description of a mechanism (Courtesy of student Joanne Fata)

when the full tube is removed, the needle stops the blood flow until another tube is punctured by the needle. This procedure can be repeated for each tube needed for specific blood tests with no discomfort to the patient. Figure 2 shows the Venoject System in a venipuncture procedure:

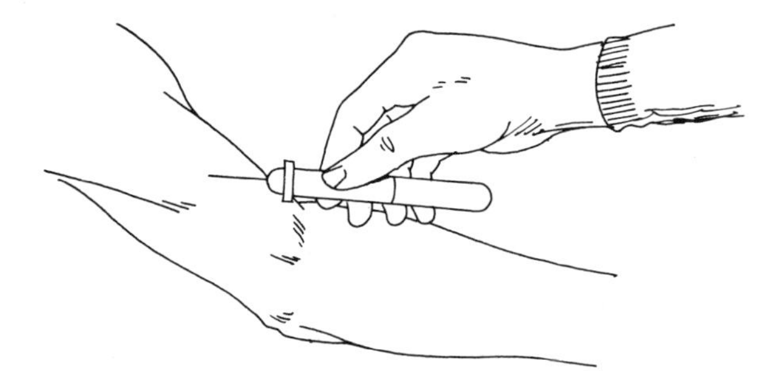

Figure 2 Venoject System in venipuncture procedure

List of Parts

The Venoject System consists of three main parts: the double-pointed needle assembly, the adapter, and the blood-collecting tube.

Functional Description

Purpose of First Part

The first main part, the needle assembly, functions in two ways: it pierces the skin at the site, and it closes off the blood flow when a collecting tube is not attached. The needle assembly consists of three subparts: the needle, the connector, and the cover. The 2.4-inch sterile needle is hollow steel. A 2.1-inch plastic connector fits securely over the needle at the halfway point. It has threads that screw into the holder and an extended tube to

Description

Figure 9.2 Continued

protect the end of the needle which is inserted into the collecting tube. A plastic cover protects the needle until it is to be used. Figure 3 illustrates the needle and plastic cover:

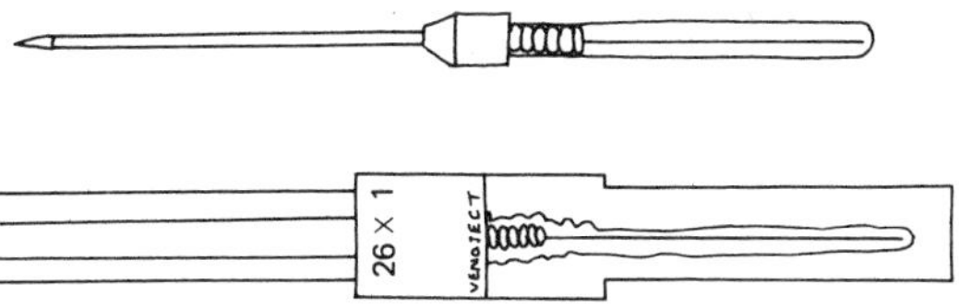

Figure 3 Venoject double-pointed needle with plastic cover

Purpose of Second Part

Description

The second functional part, the adapter holder, connects the needle and the collecting tube. The 2.8-inch, plastic, cylindrical holder has a diameter of ¾ inch. One end is threaded to receive the needle; the other end is open to receive the collecting tube. Figure 4 shows the holder:

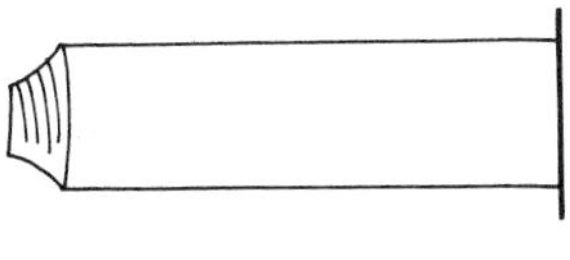

Figure 4 Venoject System holder

Figure 9.2 Continued

Purpose of Third Part

Description

The third part, the collecting tube, is designed as a vacuum to collect the blood. It consists of three subparts: the tube, a rubber stopper, and a label tape. Hollow, glass tubes are available in 3-inch, 3½-inch, and 4-inch lengths; the two shorter tubes have a ¼-inch diameter while the 4-inch tube has a ½-inch diameter. A color-coded rubber stopper is inserted into or over the open end of the tube. The color of the stopper indicates whether the tube contains an anti-coagulant which is necessary for certain blood tests. The collecting tubes in general use are not sterile; sterile tubes are available when needed for bacterial determinations. A plastic label to record the patient's name and date is taped onto every tube. Figure 5 shows three collecting tubes with rubber stoppers and labels in place:

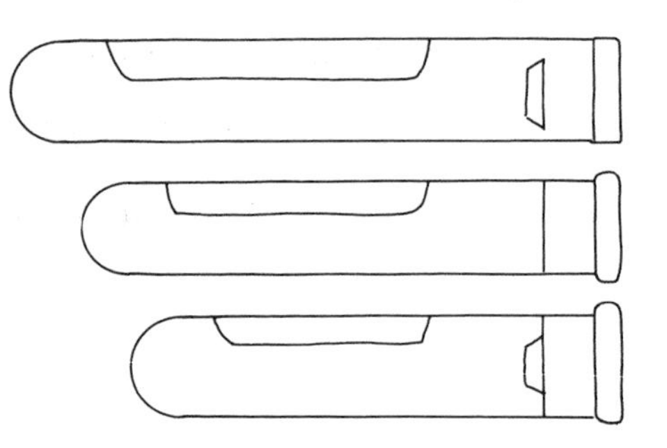

Figure 5 Venoject collecting tubes

Concluding Discussion

Assessment

The Venoject Collecting System is an efficient apparatus for collecting

Figure 9.2 Continued

blood for any number of laboratory tests. It allows the technician to obtain multiple samples while preserving the patient's vein. Because it is a disposable system, bacteria and hepatitis can not be transmitted from one patient to another. It is available at medical supply houses.

Figure 9.2 Continued

EXERCISES

1. By referring to the sample outline for a description of a mechanism in this chapter, reorganize these sentences into a logical description of an air pump.
   - **a.** The air pump consists of three main parts: barrel assembly, the plunger assembly, and the hose.
   - **b.** It is compact, lightweight, and portable.
   - **c.** The hand-operated air pump is designed for inflating bicycle tires and sporting goods, such as basketballs, footballs, rubber rafts, and so on.
   - **d.** It has a 60 psi (pounds per square inch) rating.
   - **e.** The rustproof, steel construction will ensure many years of useful service.
   - **f.** The brass, octagonal barrel cap allows access to the pump mechanism diaphragm. It is threaded to attach to the housing and has a 1/4-inch hole in its center to slide over the shaft. A 1/8-inch hole in the side of the cap allows air to enter the housing.
   - **g.** The barrel assembly consists of three subparts: the housing, a barrel cap, and a toe plate.
   - **h.** The operator clamps the hose nozzle onto the filler stem of a tire to be inflated, stands on the toe plate, and pumps the plunger up and down to inflate the tire with air. If the item to be inflated is a sporting good, the supplied filler needle is inserted into the nozzle clamp.
   - **i.** The housing is a hollow, 4 1/2-inch by 17-inch steel barrel. The top end is threaded to receive the barrel cap.
   - **j.** The 18-inch, fabric-covered, rubber hose screws into the barrel housing one inch above the base with an air-tight brass fitting.
   - **k.** A 6-inch, wood handle threads onto the top of the rod.
   - **l.** Welded to the bottom of the barrel housing is a 4 1/2-inch long toe plate base on which the operator stands during the operation of the pump's plunger mechanism.
   - **m.** The plunger assembly consists of three subparts: a rod, a handle, and a diaphragm.
   - **n.** The locking clamp nozzle is inserted into the hose end and is secured with a 1/8-inch metal band. A thumb chuck allows quick release for regular and high-pressure use.
   - **o.** The rod is a 1/4-inch by 16 1/2-inch steel shaft which is threaded at both ends.
   - **p.** The plunger assembly fits into the housing and is secured by the barrel cap.

q. The pump is 16½ inches high and is constructed of steel with rubber hosing and brass fittings.

r. A diaphragm, a leather washer, is secured to the lower end of the rod by two ¼-inch nuts.

2. Rewrite the following portion of a mechanism description to eliminate the instructional commands. Use third-person subjects, present-tense verbs in the active or passive voice.

> The operation of a socket wrench is simple. First, select the proper socket size for a specific fastener. Second, lock the socket into place on the driving lug. Third, fit the socket end of the wrench over the fastener. Fourth, set the direction control to the right or left by moving the fastener counterclockwise. Fifth, move the handle in a right to left or left to right motion to twist the fastener. Simultaneously, place your free hand over the wrench and fastener to secure the wrench to the fastener.

WRITING OPTIONS

1. Write a *general* description of one of the following mechanisms or a mechanism used in your field of study. Do not concentrate on a particular brand. Use graphics whenever possible.

a. a garlic press
b. an automobile jack
c. a drawing compass
d. a flashlight
e. a sphygmomanometer
f. a wall light switch
g. an Ohm volt meter
h. a camera
i. an alarm clock
j. an ice cream dipper
k. a manual pencil sharpener
l. a technical pen
m. needle-nose pliers
o. a tape cassette
p. a squirt gun
q. a manual can opener
r. a toy dump truck
s. a flatworm
t. the human eye
u. the human heart

2. Select a mechanism from the above list or one used in your field and write a *specific* description of it; that is, describe a specific model or brand, such as a Datsun 810 automobile jack, a Staedtler/Mars drafting compass, a Hamilton Beach, model 66, ice cream dipper. Refer to the sample outline in this chapter. Use graphics wherever possible.

10

INSTRUCTIONS

Skills:

After studying this chapter, you should be able to

1. Define *instructions.*
2. Appreciate the need for instructions in all enterprises.
3. Accurately title and provide prefatory material for a specific set of instructions.
4. Select appropriate formats for simple and complex instructions.
5. Critique a set of instructions for logical (sequential) organization and for completeness.
6. Critique a set of instructions for appropriate language.
7. Write a simple set of instructions embodying the appropriate writing strategies and graphics.
8. Write a complex set of instructions embodying the appropriate writing strategies and graphics.

INTRODUCTION

The ability to write clear, precise instructions in all career fields is crucial. Instructions are step-by-step commands which relay to the user how to locate something; how to make, assemble, maintain, or repair an item; or how to perform a task. In your professional training you have been bombarded with instructions on how to set up accounting ledgers, how to inspect a landscape project, how to administer cardiopulmonary resuscitation, how to construct a printed circuit board, how to approach a suspected felon's automobile, or how to perform other specific tasks in your particular field. In addition, you have probably mastered instructions on how to operate the mechanisms and machines used in your career.

From the first day of your employment you will rely on both oral and written instructions to learn what must be done and how to perform your tasks in the most efficient manner. Responsible employees are assigned to devise new instructions or to revise those on hand. These include how to fill out various forms, how to operate machines, how to expedite sales and services, and how to perform the myriad of processes particular to each profession.

In order to advance to supervisory and administrative positions, you, too, will need to be able to devise specific, accurate instructions for all phases of activity under your supervision.

In product-oriented businesses and industries professional technical writers combine the writing strategies of definition, descriptions of mechanisms, instructions, and sometimes analysis of a process to produce comprehensive owner and service manuals (see Chapter 8). Such writers call upon the expertise of all other employees to perfect these essential documents.

AUDIENCE

The language of instructions must be slanted toward the intended audience. Usually, the user of written instructions is a novice (a beginner); therefore, the language should be very simple and nontechnical. Even in instructions for highly technical procedures, care should be taken to define all terms which are not in general usage. Graphics are key to effective instructions. Simple procedural drawings, illustrations of the mechanisms or printed forms to be employed, or even photographs of step-by-step actions will guide the user to perform the procedure correctly.

ORGANIZATION

Titles

The first strategy in writing effective instructions is to provide a precise, specific, and limiting title. "How to Inject Lidocaine into a Finger to Produce Numbness" is far more informative than "How to Perform a Digital Block." "Instructions for Removing a Faulty Transistor from a Printed Circuit Board" is more precise and limiting than "How to Repair a Circuit Board." Your title should indicate to the reader exactly what your instructions cover.

Instructions seldom read as a continuous narrative, but are numbered to separate each command. In simple instructions an Arabic numbering system (1,2,3,4 and so on) is usually appropriate. Complex instructions may involve main steps divided into substeps. Preliminary outlining with attention to logical groupings will reveal the main steps and their parts. Several numbering systems may be considered: optional decimal systems or a digit-dash-digit system.

0.0	Preliminary section	0.0	Preliminary section
1.0	Section	1.0	Section
1.1	Component	1.01	Component

1.1.1 Subpart
1.1.2 Subpart
1.2 Component
1.2.1 Subpart
1.2.2 Subpart
2.0 Section

1.02 Component
1.021 Subpart
1.022 Subpart
2.0 Section
2.01 Component

0-1 Section
1-1 Section
    1-2 Component
    1-3 Subpart
    1-4 Subpart
1-5 Component
2-1 Section

The digit-dash-digit system relies, in part, on indentation to divide the components and subparts under the sections. The decimal numbering systems may also use indentation to aid the eye. These numbering systems are more reliable than the traditional, Roman numeral outlining system which combines Roman numerals, letters, and Arabic numerals.

Prefatory Material

Because instructions emphasize *what* to do but not *why*, it is helpful to include preliminary statements, such as naming the intended operator, stating the behavioral or instructional objective, defining key terms, or stressing the importance of the procedure. These may be included in the "O" preliminary sections of your format. Some sample prefatory materials follow:

0.0 These instructions are to be used by field service personnel to install a repaired DA-1203 antenna. Alignment of the DA-1203 to the aircraft's horizontal position gyro is necessary for proper operation of stabilized weather radar.

0-1 These instructions are intended for an allied health student learning minor surgical techniques. If followed properly, the student can anesthetize (that is, cause loss of pain and temperature sensation in) a finger. Anesthesia is necessary prior to finger surgery or before manipulation of broken finger bones. The anesthetizing of a finger or toe is termed a digital block.

Cautions, Warnings, Notes

The next strategy is to display preliminary warnings, cautions, or notes which apply to the entire procedure. Because such notations are not part of the instructions proper, it is advisable to capitalize, box, or underline such notices. Two examples follow:

CAUTION: DO NOT take an oral temperature if the patient

1. has difficulty in breathing,
2. is coughing,
3. is too weak to hold the thermometer in his or her mouth,
4. has a very dry mouth,
5. is mentally unbalanced or delirious,
6. has a nose or mouth injury,
7. is an infant or child who is not old enough to hold the thermometer in place, or
8. has had something hot or cold in the mouth within the last ten minutes.

Warning: Resist ink solvent and etching solution will cause skin damage. Always use rubber gloves when handling these substances. In case of accidental contact, wash the skin thoroughly with cold water.

Throughout your set of instructions there may be such cautions, warnings, or notes which apply to only one step. These should be placed before the step and should stand out from the actual instructional steps by the use of visuals.

Sequencing

Each of the actual steps or commands must be in chronological order. In the following excerpt from instructions on soldering a circuit board connection, Step 9 is obviously out of place; the joint would have to be secured prior to the actual soldering:

6. Preheat the joint to melt the solder.
7. Apply the solder to the joint.
8. Place the soldering iron tip against the solder and the joint for 2 or 3 seconds.
9. Secure the joint with a vise to avoid motion.

The writer of instructions is usually quite knowledgeable about the procedure, but the user is not. Therefore, not only is careful chronological order essential but also the inclusion of all steps is a must for effective operation. It is just as important to instruct users to turn on a word processing computer CRT as to instruct them on how to perform a global search of a text.

Language

Instructions demand precision, clarity, parallel construction, simplicity, and thoroughness. Each step usually begins with a command word, that is an active-voice verb stated in the imperative mood, such as *switch, disconnect, lift, depress*.

Incorrect The following tools *should be collected.*
Correct *Collect* the following tools.

Incorrect *You cut* out the premarked damaged section.
Correct *Cut* out the premarked damaged section.

Further, the command verbs should be precise.

Vague Remove the bolt.
Precise Unscrew the bolt by rotating the wrench in a counterclockwise motion.

Vague Turn on the typewriter.
Precise Depress the ON/OFF key to the ON position.

Occasionally you must precede the action command with explanatory words, such as

While depressing the RECORD button with your left forefinger, push . . .
Using a straightedge, outline the damaged portion . . .

Similarly, avoid all other vague terms.

Vague	Check the patch to ensure that a good bond has been obtained.
Precise	Pack the edges of the patch with stiff spackling compound to eliminate wobbling.
Vague	Allow the glue to dry adequately.
Precise	Allow the glue to dry for six hours.

In a set of simple instructions which are not numbered, words, such as *first, second, next, following,* and so forth, mark time and sequence.

Short sentences are easier to understand and to execute than are wordy commands.

Wordy	Insert the mounting post of the breaker arm into the recess in the cylinder so that the groove in the mounting post fits the notch in the recess of the cylinder.
Short	Fit the mounting post groove into the notch of the cylinder's recess.

Component parts of your instructions should be expressed in parallel (identical) grammatical form.

Nonparallel	1.0	Remove the damaged wall board
	2.0	Prepare the patch.
	3.0	The patch should be spackled and painted.
Parallel	1.0	Cut out the damaged wall board.
	2.0	Fit a wall-board patch into the hold.
	3.0	Spackle and paint the patch.

In your effort to eliminate wordiness, do not eliminate articles (*a, an,* or *the*) or use pronouns (*it, them, that*). Instead of writing "Push them through circuit board holes" write "Push leads A and B through the circuit board holes."

Tools, Materials, Apparatus

A command to collect or assemble the necessary tools, materials, or apparatus is logically Step 1 of most instructional sets. A list that is vague will cause frustration for the operator if he or she only realizes halfway through the task that extra tools or materials to get the job done are needed. A precise list follows which accompanies a set of instructions on how to replace the points and condenser of a Briggs and Stratton # 234 three and one-half horsepower lawnmower engine:

1.0 Collect the following tools and materials:

1.1 ¼ in. nutdrive

1.2 ⅜ in. wrench

1.3 7/16 in. wrench

1.4 flywheel holder

1.5 starter clutch wrench

1.6 Briggs and Stratton #562 flywheel puller

1.7 screwdriver

1.8 .020 in. gap tool

1.9 Briggs and Stratton #25 points and condenser with depressor tool and breaker arm spring

1.10 flywheel key

1.11 5 or 6 large rags

List such obvious items as old newspapers, running water, pencils, and paper. Do not assume that the user of your instructions has any familiarity with the task that you have probably performed many times.

Figures 10.1 and 10.2 show sample sets of instructions. Figure 10.1 uses a simple Arabic numbering system and graphics. Figure 10.2, because it involves main and substeps, illustrates a complex numbering system. Be prepared to critique each set in class.

HOW TO ASSEMBLE A HANDLEBAR AND STEM WITH A LOCKING WEDGE ON A BICYCLE

To prevent any instructions or parts from being discarded, keep the carton until the handlebar is completely assembled. Keep all instructions on assembly for future reference.

NOTE: Torque (the measurement which indicates how much a nut or bolt must be tightened) is measured in foot pounds with a torque wrench.

1. Collect the following tools:

   adjustable wrench

   torque wrench

2. Slide the expander bolt washer onto the expander bolt.
3. Insert the bolt with washer into the top of the stem.
4. Slide the wedge onto the protruding bottom of the expander bolt as shown.
5. Unscrew counterclockwise the clamp nut, washer, and the

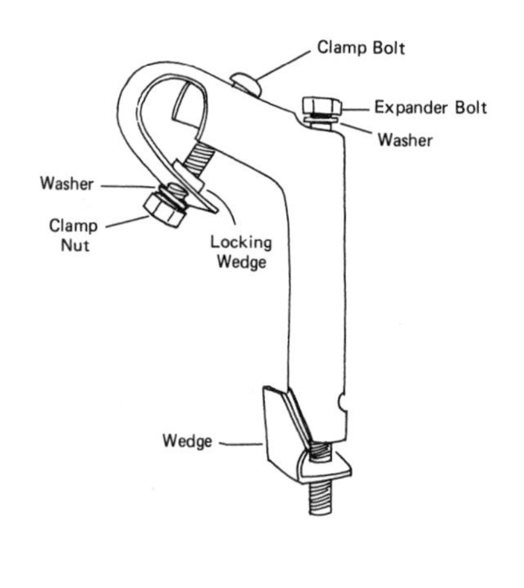

Figure 10.1 Sample simple instructions

locking wedge and remove the part from the stem.

6. Fit the stem over the small curved area of the handlebar and slide the stem to the center of the handlebar as shown.
7. Assemble the clamp bolt, locking wedge, washer, and the clamp nut to the stem as shown and tighten loosely.
8. Push the stem into the fork tube at least 2½ in. as shown on the side of the stem.
9. Align the straight part of the handlebar at a right angle (90 degrees) to the front wheel as shown.
10. Using a torque wrench, tighten the expander bolt to a torque of 18 foot pounds.

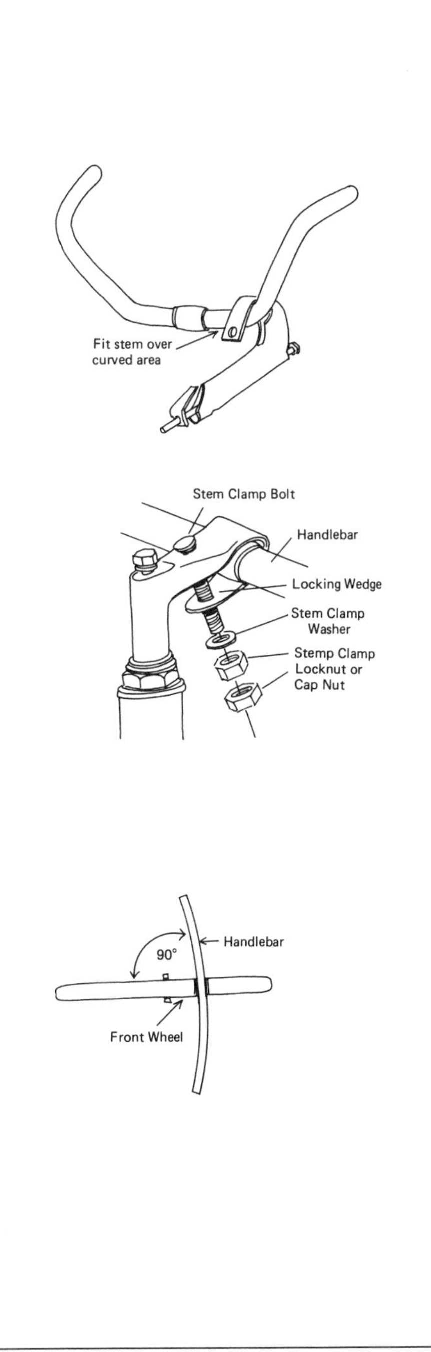

Figure 10.1 Continued

CAUTION: APPLY THE CORRECT TORQUE TO THE EXPANDER BOLT OR THE FORK TUBE OF THE BICYCLE CAN BE DAMAGED CAUSING A PROBLEM WITH THE STEERING.

11. Making sure the stem is at the center of the handlebar, use a torque wrench to tighten the clamp nut to a torque of 23 foot pounds.

Figure 10.1 Continued

INSTRUCTIONS FOR TAKING AN ORAL
TEMPERATURE WITH AN ELECTRO:THERM

0.0 These instructions are intended for a medical assistant who is working in a doctor's office. If followed properly, the assistant will obtain an accurate temperature reading in less time than if one used a mercury type thermometer.

> CAUTION: DO NOT take an oral temperature with the electro:therm if the patient
> 1. has a mouth injury,
> 2. is an infant or young child who is not old enough to hold his lips closed when told to do so, or
> 3. has had something hot or cold in the mouth in the last five minutes.

1.0 Collect the following apparatus and place on the counter:
- 1.1 an electro:therm thermometer
- 1.2 a box of sterile plastic covers that are made to be used with the electro:therm
- 1.3 a box of clean tissues

2.0 Familiarize yourself with the electro:therm by studying the parts as diagrammed in Figure 1:

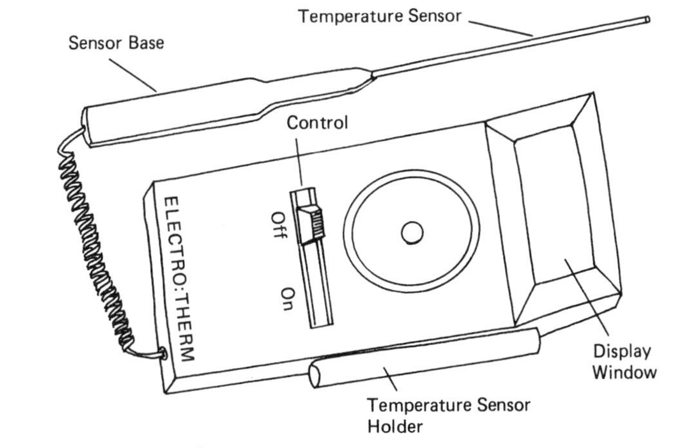

Figure 1. Parts of an Electro:therm

Figure 10.2 Sample instructions with complex numbering system

3.0 Prepare the electro:therm for operation.
- 3.1 Remove the temperature sensor from the control base by pulling the temperature sensor back towards the cord with your left thumb and first finger.
- 3.2 With your right hand, pick up one sterile plastic cover by its paper wrapping.

WARNING: Your hands should NOT come in contact with the sterile plastic covering.

- 3.3 Insert the temperature sensor into the sterile plastic cover until the plastic cover is completely engaged over the narrow sensor rod.
- 3.4 Gently pull off the paper wrapping leaving the sterile plastic cover on the temperature sensor.

Figure 2 shows what the temperature sensor looks like with the plastic in place:

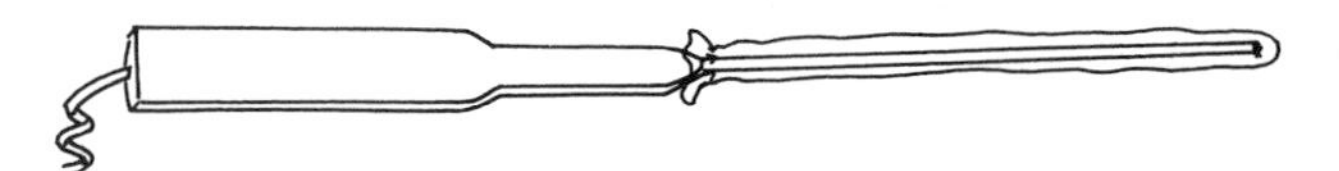

Figure 2. Temperature sensor with plastic cover in place.

3.5 Switch the base of the temperature sensor to your right thumb and first finger.
3.6 Pick up the control base with your left hand.
3.7 With your left thumb, turn the control base to the ON position.
3.8 Observe the electro:therm's flashing numbers in the display window.

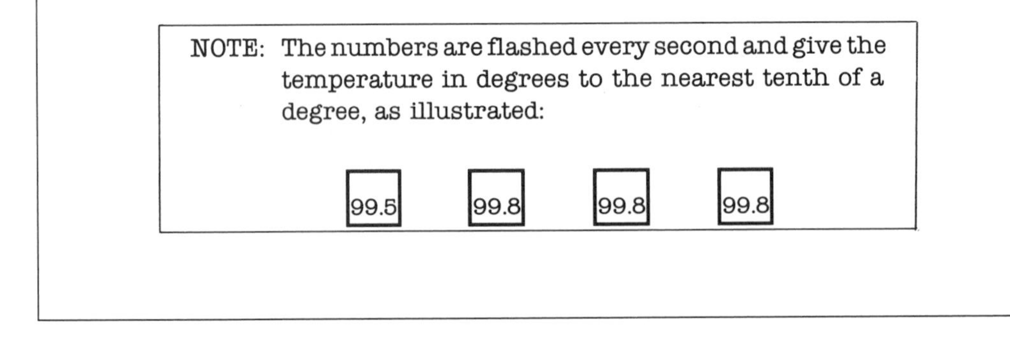
NOTE: The numbers are flashed every second and give the temperature in degrees to the nearest tenth of a degree, as illustrated:

Figure 10.2 Continued

4.0 Take temperature reading.

CAUTION:	Unlike the mercury thermometer, the temperature sensor should be held continually in the patient's mouth while taking the temperature. DO NOT remove the sensor from the patient's mouth until the same degree displays at least two times in a row. The electro:therm is so sensitive that a patient taking a breath can lower the reading.

4.1 Insert the sterile, plastic-covered temperature sensor into the patient's mouth making sure it is under the tongue.
4.2 Instruct the patient to keep lips closed tightly and not to talk.
4.3 Leave the temperature sensor in the patient's mouth for one minute.
4.4 After the minute has lapsed, observe the temperature as displayed on the control base.
4.5 When the same numbers have flashed for two times, remove the temperature sensor from the patient's mouth.

5.0 Deactivate the electro:therm.
5.1 Turn the control base switch to the OFF position with your left thumb.
5.2 Place the control base on the counter.
5.3 Switch the base of the temperature sensor to your left thumb and first finger.
5.4 With a tissue in your right hand, pull the used plastic cover off of the temperature sensor and discard the tissue and the plastic cover.
5.5 Slide the temperature sensor back into its holder on the side of the contrl base.

6.0 Record the patient's temperature on the chart.

Figure 10.2 Continued

EXERCISES

1. Most of you have recorded a cassette tape and are familiar with the recording steps. Edit and rewrite the following instructions. Look for errors in organization, sequence, and completeness. Remember that notes, warnings, and cautions should be separated from the actual steps.

INSTRUCTIONS FOR RECORDING CASSETTE TAPES
ON A STEREO SYSTEM

1. Press the PAUSE key.
2. Turn the selector to the program source: AM STEREO, FM STEREO, PHONO, OR AUXILIARY.
3. Press the STOP/EJECT key to open the tape door.
4. Simultaneously press the PLAY and RECORD keys.
5. Adjust the Record Level Controls so that the left and right VU meters will read −3 to 0.
6. VU meters show the recording and relative playback level of each channel. During recording, the meter pointer should move as high as possible, but not past the 0 db mark for best sound.
7. When ready to record, press the PAUSE key again to start the tape movement.
8. When one side is recorded, the tape will automatically stop.
9. Press the STOP/EJECT key again to open the door to remove the tape.
10. To stop recording, press the STOP/EJECT key once.

2. Edit the following set of instructions for language problems. Look for verbs which are not expressed in the imperative mood, active voice. Also look for vague verbs and other terms, too lengthy sentences, nonparallel constructions, eliminated articles, and questionable pronoun references.

INSTRUCTIONS FOR THE HOLGER-NIELSON
(BACK PRESSURE-ARM LIFT) METHOD OF
MANUAL ARTIFICIAL RESPIRATION

1-1 Positioning the victim

1-2 Check the victim/s mouth for foreign matter and wipe it out quickly.

1-3 WARNING: CHECK it for obstructions every 30 seconds.

1-4 Place the victim face down, bend elbows, and hands are upon the other.

1-5 The victim's head should be turned slightly to one side with head extended and chin jutting out.

2-1 Administering the respiration
    2-2 Kneel at his head.
    2-3 You place your hands on the flat of victim's back.
    2-4 Rock forward until your arms are vertical to his back.
    2-5 The weight of the upper part of your body should now be forced down to exert a steady, even pressure downward upon your hands which are already placed on the back of the victim.
    2-6 Slide your arms to the arms of the victim just above his elbows and draw the arms upward and toward you.
    2-7 NOTE: Enough lift to feel resistance and tension in his shoulders should be applied.
    2-8 The victim's arm must be lowered to the ground.
    2-9 Repeat this cycle 12 times a minute.

WRITING OPTIONS

1. Write a short (12–15 steps) set of instructions (using a simple, Arabic numbering system) for one of the following subjects. Use graphics.
   a. How to clean and insert extended wear contact lenses
   b. How to take a telephone message at your place of employment
   c. How to take an oral temperature
   d. How to use a microwave corn popper
   e. How to open a manuscript on a specific word processing computer
   f. How to make a thin blood smear
   g. How to change a tire on a specific model car
   h. How to use a snorkel mask
   i. How to reconcile a bank statement
   j. How to vacuum a rectangular swimming pool
   k. How to defrost a steak in a specific model microwave oven
   l. How to open a bottle with a Vintage cork puller
   m. How to load film into a specific make camera
   n. How to start a flooded car
   o. How to rescreen a door
   p. How to make a magazine rack
   q. How to operate an electric sander (or any other power tool)
   r. How to administer CPR
   s. How to iron a dress shirt or blouse
2. Write a complex set of instructions which involves main steps and substeps. Select a decimal or digit-dash-digit numbering system. Select a subject from the following list or use a subject from your career field. Use graphics.

- **a.** How to perform a preflight aircraft check
- **b.** How to insert, delete, and move block data on a specific word processing computer
- **c.** How to inspect a building for fire safety (plumbing, electrical, or other construction detail)
- **d.** How to administer glucose intravenously
- **e.** How to propagate a plant for air-layering
- **f.** How to reweb lawn or beach furniture
- **g.** How to install vertical blinds (floor tile, accoustical tile, and so on)
- **h.** How to operate a scuba regulator
- **i.** How to fill out an expense report at your place of employment
- **j.** How to make a printed circuit board
- **k.** How to perform any procedure on a specific computer
- **l.** How to perform any marketing procedure
- **m.** How to perform any aviation technology procedure
- **n.** How to perform any electronic technology procedure
- **o.** How to perform any allied health procedure
- **p.** How to perform any technical procedure in your field

3. Write a detailed set of instructions to a hypothetical substitute who must perform your job any given day next week.

11

PROCESS ANALYSIS

Skills:

After studying this chapter, you should be able to

1. Understand the purpose of process analysis.
2. Identify three types of process analysis.
3. Distinguish between a process analysis and a set of instructions.
4. Name the logical divisions of a process analysis.
5. Write an historical, scientific, mechanical; or natural; or organizational process analysis embodying logical organization, appropriate language, and graphics.

INTRODUCTION

Process analysis is a method of explaining how something occurred, how something is done, or how something is organized by separating the process into its parts to find out their nature, proportion, function, and relationship.

A process analysis may contain elements similar to instructions or mechanism descriptions but should not be confused with either. Instructions emphasize *what to do,* and mechanism descriptions emphasize *how something is put together;* neither often explains *why.* An effective analysis of a process emphasizes *what, where, when, how, to what degree, to what extent, under what conditions,* and, most important of all, *why.* The reader of an analysis must be able to judge the reliability, practicality, and efficiency of the process. The reader should be able to estimate the difficulty of the process, the problems likely to occur, and successful solutions to these problems.

Types of Analysis

An informational process analysis informs the reader about a particular process for the purpose of increasing his or her general knowledge. There are three basic types of process analysis:

1. The historical analysis explains how and why an idea, event, or institution occurred.
2. The scientific, mechanical, or natural analysis explains how such processes occur or should occur.

3. The organizational analysis explains the steps, pitfalls, and methods of efficiently performing a human process.

Historical analysis explores such subjects as how the microcomputer was developed or how American teachers unionized. Subjects appropriate to scientific, mechanical, or natural analysis include how chemotherapy cures cancer, how sound waves are transmitted, or how the eye sees. Organizational analysis examines subjects such as how a plant is propogated or how a manager motivates a staff of workers.

Audience

The reader of any given process analysis is usually a novice; therefore, the language should not be highly technical. The writer should provide ample background material, define all terms, and use simple language. Remember that the reader wants to judge the process. Graphics, such as a flow chart or pictograph of the steps, drawings, or schematics of the process steps, and charts or tables of costs and time are helpful.

Language

Directions are written in the imperative mood to give a series of commands: *put, spread, you apply,* and so on. Process analyses are written in the indicative mood to explain steps: the technician *applies,* the worker *spreads,* the layer of resin *is smoothed.* These indicative voice verbs may be active or passive.

Avoid the words *you* and *your* in a process analysis.

Incorrect All banks recommend when *you* receive *your* monthly bank statement that *you* reconcile *your* records immediately.

Correct All banks recommend that their depositors reconcile their records as soon as they receive their monthly bank statements.

or

All banks recommend checking the monthly statements immediately to rectify personal records.

ORGANIZATION AND FORMAT

A process analysis essentially consists of three parts: prefatory material, an analysis of the steps, and concluding discussion. This organization is similar to that used in a description of a mechanism (see Chapter 9). The following presents a typical process analysis.

1.0 Prefatory material (introduction)
- **1.1** Intended audience
- **1.2** Definition/purpose
- **1.3** Background and theory
- **1.4** Who, when, where
- **1.5** Special considerations
- **1.6** Tools, materials, supplies, apparatus, (if applicable with graphics)
- **1.7** List of chronological steps (or graphic flow chart)

2.0 Analysis of the steps (body)
- **2.1** First main step (with appropriate graphics)
  - **2.1.1** Definition and/or purpose
  - **2.1.2** Theory (if applicable only to the specific step)
  - **2.1.3** Special considerations (if applicable only to the specific step)
  - **2.1.4** Substeps (if applicable)
  - **2.1.5** Analysis of the step
- **2.2** Second main step . . . (etc.)

3.1 Concluding discussion/assessment (closing)
- **3.1** Results evaluation
- **3.2** Time and costs (if applicable)
- **3.3** Advantages
- **3.4** Disadvantages
- **3.5** Effectiveness
- **3.6** Importance
- **3.7** Relationship to larger process

A process analysis is usually written in a narrative format although it may employ outline, decimal numbering, or digit-dash-digit techniques (see Chapter 10). The writing strategies of process analysis may be interspersed with instructions and descriptions of mechanisms in an operator's or service manual (see the sample in Chapter 7).

PREFATORY MATERIAL

Titles

The title should be precise, descriptive, and limiting. Avoid a "how-to" title which implies that instructions, rather than an analysis, are to follow.

Examples

The Process of Taking Blood Pressure with a Sphygmomanometer

Preparing an Income Statement for a Small Business

How a Congressional Bill Becomes a Law

Audience Statement

You must consider the purpose and audience of your process analysis. Your audience will determine the extent of your analysis and the degree of technicality in it. State exactly what the purpose is and for whom the analysis is intended.

Examples

The process analysis of taking a patient's blood pressure is designed for a beginning nursing student with no prior experience

This analysis of fire code inspection procedures, for fire chiefs, firefighters, and fire fighter trainees, is designed to simplify and to regularize practices.

Definition/Purpose

A formal definition of the process or a clear statement of the purpose of the process is necessary before an analysis of the parts. The prefatory material may include definitions of other key terms to be used within the body of the report to avoid later interruption.

Example

Blood pressure is the force exerted by the blood against the walls of the blood vessels. It is created by the pumping action of the heart. The greatest pressure, known as *systolic pressure,* occurs during the contraction of the heart. The lowest pressure, known as *diastolic pressure,* occurs during the relaxation or rest period of the heart. The purpose of taking a patient's blood pressure is to relate it to other health factors, to determine if the patient is healthy, or to determine the cause of illness.

Background/Theory

It may be necessary to explain the historical or scientific background, theory, or principle of a process. Such discussion also belongs in the prefatory material.

Example

A brief understanding of the conditions under which blood circulates in the body is necessary. Blood passes from the heart throughout the body by way of a system of vessels that eventually return the blood to the heart. This journey is so rapid that a single drop of blood usually requires less than one minute to move from the heart through the body and back to the heart.

A single tube leading from the heart divides into smaller and smaller vessels, the arteries. The smallest arteries branch into capillaries, the most minute blood vessels. Through the thin walls of the capillaries, the blood supplies the body with oxygen from the lungs and collects the waste products of the body for subsequent removal by the kidneys and other excretory channels.

Beyond the capillaries, the branching process is reversed. The capillaries join to make slightly larger tubes, which next unite to form larger vessels, the veins. Eventually the blood is returned to the heart by two large veins. Therefore, pressure is greatest in the arteries and least in the veins.

One records blood pressure as a fraction, such as 120/80 mm Hg (the chemical symbol for mercury). The normal blood pressure for a healthy, resting adult ranges from 100 to 140 mm Hg systolic and from 60 to 90 mm Hg diastolic.

Who/When/Where

If your analysis involves an operator, the qualifications of the operator should be specified in the prefatory material. When and where the process is performed should also be explained.

Example

Nurses, doctors, medical assistants, and other paraprofessionals trained in the use of a sphygmomanometer determine blood pressure in the daily routine of patient care for diagnostic purposes. Because blood pressure will vary at different times of the day and because readings are usually taken for the purpose of comparison with previous readings, readings should be taken at the same time every day. Generally, only a doctor is qualified to evaluate blood pressure readings in relation to a patient's sickness or health. The readings are taken in a doctor's office, a hospital, at the scene of medical rescue operations, or in other clinical settings.

Special Considerations

Special conditions, requirements, preparations, and precautions which pertain to the entire process, and not just to one step, should be indicated in the prefatory material.

Example

It should be noted that many factors influence blood pressure. A patient should be asked if any of the following factors could be influencing his blood pressure at the time of the reading:

1. Increasing blood pressure factors
   a. eating
   b. stimulants
   c. exercise
   d. emotional stress
2. Decreasing blood pressure factors
   a. rest
   b. fasting
   c. depression

Other factors which should be considered are pain, climate variation, tobacco, bladder distension, hemorrhage (blood loss), blood viscosity (thickness), and the elasticity of the arteries.

Tools/Materials/Supplies/Apparatus

A precise, detailed list of all tools, materials, supplies, and apparatus used in the performance of the process should be provided. It may be necessary to write a brief description of an unusual mechanism used in the process (see Chapter 9).

Example

The following supplies are used to take an accurate blood pressure reading:

1. Stethoscope, an instrument used to magnify the sounds of arterial pulse.
2. Sphygmomanometer, a three-part instrument consisting of a mercury pressure gauge, an arm band with an inflatable rubber bladder, and a pressure bulb to control the flow of air going through connecting tubes in and out of the bladder.
3. Alcohol to clean the earpieces of the stethoscope.
4. Cotton balls to apply the alcohol.

List of Chronological Steps

After dividing the process into its main steps, each based on completion of a stage of work or action, the steps should be listed in chronological sequence: first, second, third, and so on. Use the following form or a flow chart.

Pattern The process consists of ______________ main steps:
first, ______________ ing the ______________;
second, ______________ ing the ______________;

third, ________________ing the ________________;
and finally, ________________ing the ________________.

Example The process of taking a blood pressure reading consists of nine main steps: first, preparing the patient; second, assembling the sphygmomanometer; third, attaching the arm cuff to the patient; fourth, placing the stethoscope over the brachial artery; fifth, closing the pressure control valve; sixth, inflating the cuff; seventh, opening the control valve; eighth, taking the reading; and ninth, removing the apparatus.

ANALYSIS OF THE STEPS

The body of the process analysis thoroughly examines and analyzes each step. Your emphasis here should be to explain why the process is performed in a particular manner. Analyze what would happen if the process were not performed in the correct manner.

Definition/Purpose

Identify each step and write a formal definition and/or a statement of purpose of the step.

Example

The fourth step, placing the stethoscope over the brachial artery, is done so that the clinician can hear the rhythmical, thumping sounds of the blood. The brachial artery is the large artery of the arm at the inner crease of the elbow.

Theory

If a particular step is based on a theory, explain how the theory applies to the step.

Example

The brachial artery is used because it is near the heart, large enough for specific recognition, and near the surface of the skin.

Special Conditions

Describe in detail any special considerations, requirements, apparatus, preparations, and precautions which apply to this step only.

Example

In most patients, the brachial artery is found quite simply. If there is any difficulty in locating it, the opposite arm may be more yielding. An injured arm or one that contains an intraveneous injection should not be used.

Substeps

If a main step contains substeps, list them chronologically prior to explaining each.

Example

The third step, attaching the arm cuff to the patient, consists of three substeps: checking the cuff bladder, positioning the cuff, and attaching the cuff to the arm.

Analysis of Steps

Finally, explain each step and its substeps with attention to the reasons each is performed in a specific manner.

Example

The cuff bladder should not contain any air at the time of positioning the cuff. If it does, a secure fit will not be affected. The armband is wound around the arm above the elbow at a level with the heart allowing room beneath it for the stethoscope bell. The band is fastened by means of the hooks, snaps, or Velcro material provided for this purpose. If no means of fastening are provided, the end of the band may be tucked securely under the top of the band. If the band is wound too tightly, it will bind the arm, create extra pressure, and cause discomfort to the patient. If the band is wound too loosely, the sounds will be deadened by the cushion of air required to tighten the band sufficiently to compress the brachial artery.

Each main chronological step should be analyzed by the same writing strategies.

CONCLUDING DISCUSSION/ASSESSMENT

One or more of the following factors should be discussed in the conclusion:

- Time and cost
- Advantages
- Disadvantages
- Effectiveness
- Importance
- Relationship to a larger process

Some of these points may already have been covered in the introduction. Keep in mind that your reader is seeking to assess the efficiency of the process.

The following concluding discussion explains time and cost, advantages, importance, and relationship to the process.

Example

Taking a blood pressure reading requires only a few minutes. Because such readings are routine in regular medical check-ups and patient care, the cost is included in the consultation fee. Many health associations provide free blood pressure readings to the public in such places as shopping centers, libraries, and schools.

A blood pressure reading is the one sure method of detecting hypertension, the silent killer. Early detection can prevent strokes, coronaries, and kidney failures.

It is important to have an ongoing record of readings so that variations can be detected. By itself, a reading will not tell the doctor what is wrong, but along with other diagnostic procedures, it will help to determine a patient's condition.

Figure 11.1 shows another organizational process analysis. Be prepared to critique the sample in class.

STOPPING THE FELONY SUSPECT'S VEHICLE

Audience

This analysis is designed for police academy trainees with no prior police experiences.

Purpose/Definition

The purpose of this analysis is to familiarize police academy students with the correct, efficient, and safe way to stop a felony suspect in a vehicle. A felony is any crime, defined by law, for which the culprit could receive imprisonment in a penitentiary or the death sentence.

Background

During a tour of duty a police officer may observe thousands of vehicles. He may observe a license number of a wanted or stolen vehicle, or he may recognize a wanted criminal inside a vehicle. In any such encounter, the police officer must immediately distinguish between proper procedures and carelessness. At times, officers have failed to make full use of the advantages which proper procedures offer. Fear of being called "overcautious" of or "crying wolf" have led some officers to disregard the responsibility of self protection.

Precaution

Before stopping a suspected felon in a vehicle, the officer should ascertain that a supporting or back-up officer will be available for assistance.

Who/When/Where

This process could be performed by any police officer in any city, county, or state when a felony suspect is spotted in a vehicle.

Tools

The officer requires a patrol car, a two-way radio, and a service pistol.

Steps

Stopping a felony suspect in a vehicle consists of five main steps:

Figure 11.1 Sample process analysis

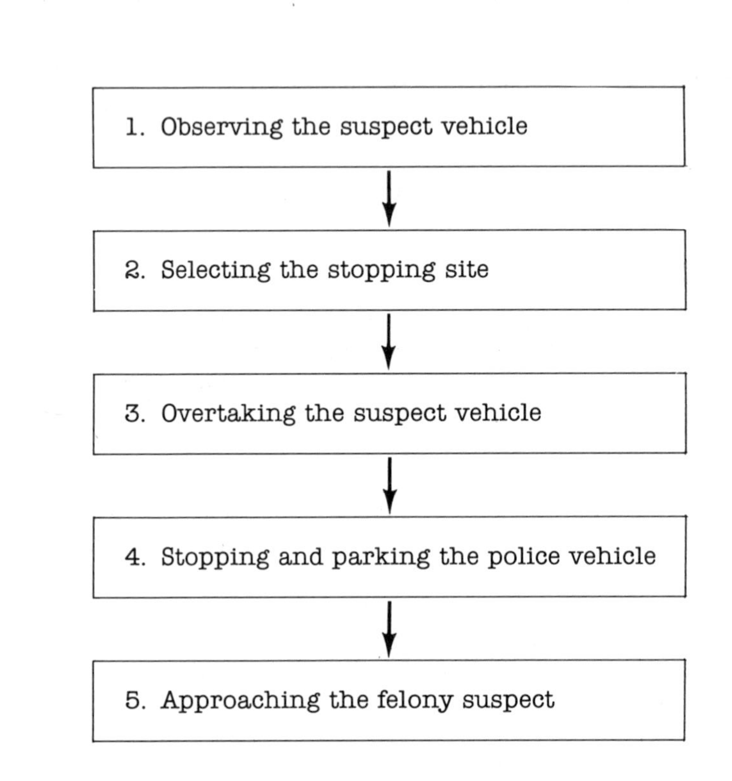

Analysis of Steps

The first step, observing the suspect's vehicle, requires the officer to notify the radio dispatcher at once. The officer gives the following information:

1. identification of the police unit,
2. location of the contact,
3. description of the vehicle and license number,
4. description and number of occupants, and
5. direction of travel and name of the last cross street passed.

The last cross street is broadcast at intervals to aid the dispatcher in predicting the suspect's course of travel, thereby hastening the arrival of supporting police units. The officer writes the license number, description of the car, and the number of occupants on a memo pad inside the police car. This is necessary in case the dispatcher did not receive all of the officer's radio transmission. The suspect's vehicle is trailed until the supporting

Figure 11.1 Continued

units are close. The officer should be alert for sudden stops, turns, or other evasive action on the part of the suspect's vehicle.

The second step, selecting a stopping site, requires the officer to select a location with which he is familiar. Familiar surroundings will be to the officer's advantage as he will be able to direct additional assistance to the location more quickly. The officer will also be in a better position to make an apprehension if the suspect attempts to flee. Stopping near alley entrances, openings between buildings, vacant lots, and other easy escape should be avoided. At night, a well-lighted area will enable the officer to observe whether or not the suspect is disposing of any evidence or a weapon.

In the third step, overtaking the suspect's vehicle, the officer maintains constant vigilance to guard against any evasive action on the part of the suspect. At this time, the officer operates the siren and vehicle emergency lights. The police car is driven directly to the rear of the suspect's vehicle. Overtaking the suspect's vehicle is illustrated in Figure 1:

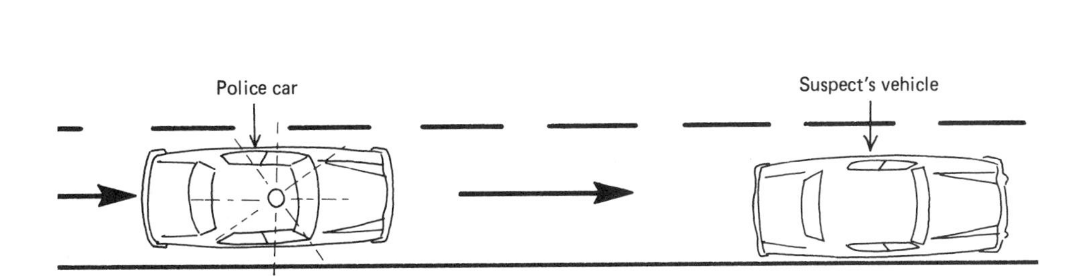

Figure 1 Overtaking the suspect's vehicle

The distance between the two vehicles will vary with the speed. Usually a distance of one car length for every ten miles per hour provides a safety zone for the officer. The officer should constantly consider the safety of other motorists and pedestrians in overtaking and stopping the suspect's vehicle. The suspect's vehicle should be stopped as far as possible to the right side of the roadway or off of the roadway altogether.

In the fourth step, stopping and parking the police car, the officer notifies the dispatcher of the location of the stop. The officer should park his vehicle about 10 feet behind the suspect's vehicle with the front of the police vehicle pointing toward the center of the street. The police vehicle should be on a 45° angle to the suspect's vehicle. The police vehicle position is illustrated in Figure 2:

Figure 11.1 Continued

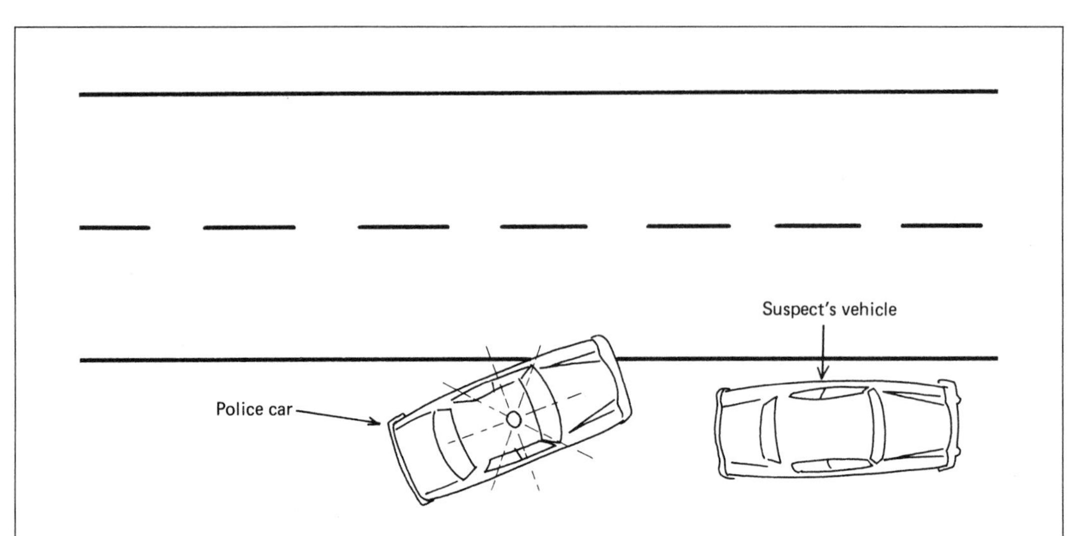

Figure 2 Parking position of stopped vehicles

The police vehicle emergency lights are left on so that other motorists will be warned of danger and the supporting police units will be able to find the officer quickly.

In the fifth and final step, approaching the suspect, the officer making the stop should take complete command of the situation. He should get out of the police vehicle and with gun drawn proceed to the left front fender of the police vehicle. Using the body and engine of the vehicle as protection, the officer identifies himself in a loud, clear voice: "Police officer! You are under arrest! Turn off your motor and drop the keys on the ground." The officer may order the suspect to place both of his hands out the driver's window or to place his palms against the inside of the windshield. The officer should stay by his police vehicle until a supporting or back-up officer arrives. The back-up officer should park his police vehicle to the rear and a little to the right of the first police car.The back-up officer should proceed to a position off the right rear of the suspect's vehicle with his gun drawn. The position of the police vehicles and officers are illustrated in Figure 3:

Figure 11.1 Continued

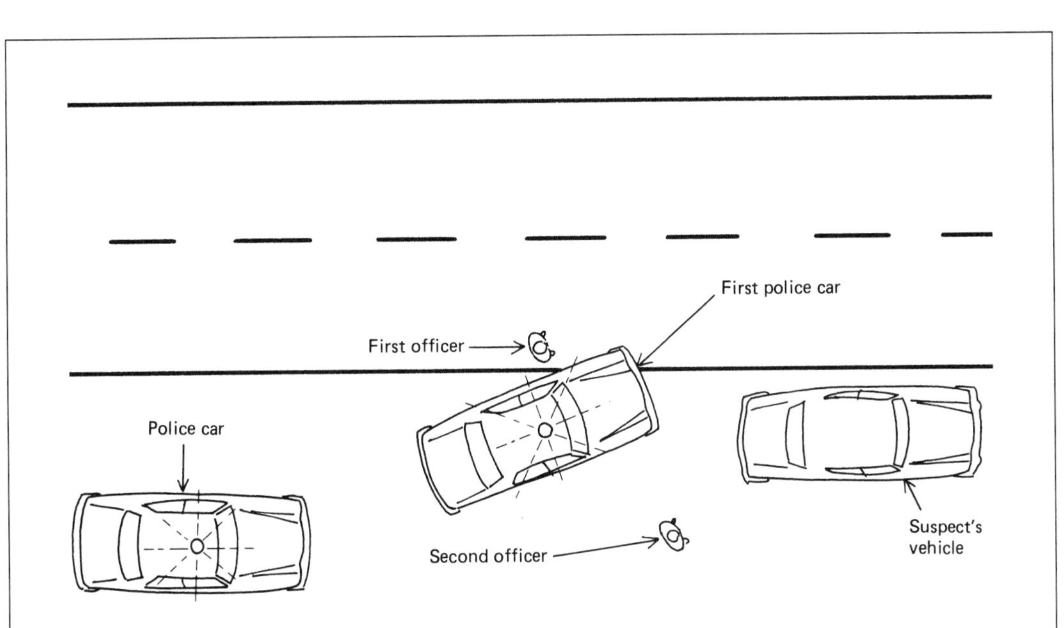

Figure 3 Position of vehicles and officers

The suspect should be made aware of the second officer's presence by the first officer to reduce the possibility of sudden attack from the suspect. The suspect is ordered out of his vehicle by the first officer. The first officer then orders the suspect to assume a search position.

Conclusion

Searching a suspect is another process and is not discussed here. No two felony stops are identical in nature. The unpredictable actions of the felony suspect make the officer rely on his training and judgment in each case. By following the steps in this process, the danger to the officer is minimized.

Figure 11.1 Continued

EXERCISES

1. Classify the following process analysis subjects as historical (H); scientific, mechanical, or natural (S); or organizational (O).
   a. How Volcanoes Occur
   b. Changing a Tire on a Hill
   c. How Kidneys Function
   d. Planning a Trip to Spain
   e. The Construction of Disneyland
   f. How Herpes Simplex II Is Transmitted
   g. How Alaska Became a State
   h. How Intravenous Fluids Are Administered
2. The following excerpt is from a process analysis of developing black and white film. Rewrite it to eliminate commands and *you* and *your*. Use third person subjects and indicative mood verbs in the active or passive voice.

   In the first step, loading the film into the roll, you attach the film to the developing reel. Place the end of the film into the notch in the reel. Wind the entire roll onto your reel. Next, you place the reel into the developing tank. Snap the top securely into place. You may now turn on your lights.

WRITING OPTIONS

1. Write a process analysis employing logical organization, appropriate language, and graphics. You may select a subject from the following list or analyze a process performed in your particular field of study.
   1. How Women Won the Vote
   2. How the Ford Mustang Has Been Modified
   3. How China Became Communist
   4. How the Rocky Mountains Evolved
   5. How Microcomputers Have Been Improved in Ten Years
   6. How Abortions Became Legal
   7. How Legionnaires Disease Was Investigated
   8. How Astronauts Are Trained
   9. Tuning a Motor of a Specific Car
   10. Taking X-Rays of the Hand
   11. Preparing a Porcelain Filling
   12. Performing a Root Canal

13. Blowing Glass
14. Preparing a Stencil
15. How Gum Disease Develops
16. How an Earthquake Occurs
17. How a Television Image Is Transmitted
18. How the Lungs Function
19. How Soap Is Manufactured
20. How Osmosis Occurs
21. Applying Vinyl Tiles
22. Editing a Manuscript on a Word Processing Computer
23. Selecting an Automobile
24. Judging a Debate
25. Reconciling a Checkbook
26. Planning and Constructing a Rock Garden
27. Redecorating a Room
28. Taking Fingerprints
29. Preparing a Payroll Summary Sheet
30. Performing a Preflight Check
31. Landing a Single Engine Airplane
32. Writing an Airline Toll Ticket
33. Apprehending a Shoplifter
34. Administering Intraveneous Fluids
35. Inserting a Foley Catheter
36. Removing Sutures
37. Applying a Dressing to a Wound
38. Checking a Credit Rating
39. Establishing an Equal Opportunity Program
40. Interviewing a Job Candidate

12

RESEARCH AND DOCUMENTED REPORTS

Skills:

After studying this chapter, you should be able to

1. Understand the processes of research and documentation
2. List the strategies involved in researching and documenting a report.
3. Appreciate the importance of selecting a limited subject for research.
4. Formulate a thesis to organize research and focus the purpose of a report.
5. Access material from the various catalogues, reference books, indexes, abstracts, and files in a library.
6. Prepare bibliography cards and listings appropriate to MLA, Author/Date, and Number systems.
7. Organize a report by various writing strategies.
8. Appreciate and demonstrate preliminary and final outline strategies.
9. Take summary, paraphrase, and quotation notes that are relevant to a report and free of plagiarism.
10. Edit a rough and final draft of a research report.
11. Provide the appropriate documentation for a research report.

INTRODUCTION

Research is the process of investigating and discovering facts. Research may consist of testing, polls, questionnaires, interviews, letters of inquiry, observation, and the like, or it may consist of investigating the works of other researchers who have published in general books, encyclopedias, abstracts, specialized dictionaries, handbooks, manuals, almanacs, pamphlets, and so on. A research report gives formal credit (documentation) to the sources used or quoted; that is, parenthetical references to the works cited are included within the text, and a list of "Works Cited" is included at the end.

The research or documented report involves ten strategies:

Strategy 1. Selecting and limiting the subject

Strategy 2. Formulating the thesis

Strategy 3. Researching the source material

Strategy 4. Preparing a working bibliography

Strategy 5. Brainstorming the writing strategies

Strategy 6. Preparing the working outline

Strategy 7. Taking notes

Strategy 8. Writing the rough draft

Strategy 9. Referencing the sources

Strategy 10. Preparing the final draft

STRATEGY 1. SELECTING AND LIMITING THE SUBJECT

Sometimes a supervisor or professor assigns you a research report topic, such as "The Influence of Temperature on Viruses," "The Effects of the Beta Blockers on Coronary Patients," or "The Impact of the Microchip on Credit Cards." On the other hand, the selection of the subject may be yours. In selecting a topic for research, three principles must be considered:

1. Select a subject which can be thoroughly investigated within the confines of your length limitations. "Computers" is too vast a subject and fails to suggest a particular focus. On the other hand, "The Effectiveness of Computer Access Control Devices" is both more narrowed and focused.
2. Select a subject on which a wide variety of research material is published. "The Causes of Acquired Immune Deficiency Syndrome (AIDS)" will not be productive because research about the causes is too recent and inconclusive. "The Effects of Nifedipine" may be too specialized for you and your audience. "The Restoration Problems of Archeologists Excavating the Xian (China) Ruins" is too far removed for you to obtain published, up-to-date materials.
3. Select a subject which will allow you to formulate a judgment. A study on "The Use of Computers" will yield plentiful material, but a study of "The Effects of Computers on Employment Levels in the Accounting Field" will yield information on which you can formulate an informed opinion.

STRATEGY 2. FORMULATING THE THESIS

The next strategy is to develop a preliminary thesis, a statement which focuses the purpose of your paper. This statement of the central idea to be developed in your paper will help you to organize facts, limit your note-taking, and eliminate needless research.

Consider what you know about your limited subject. You probably already have some opinions or you would not have selected the subject in the first place. State in one or two sentences an opinion, conclusion, generalization, or prediction about your subject. For instance, if you have narrowed your subject to "The Effectiveness of Computer Access Control Systems," you may tentatively state, "Computer access control systems are inadequate."

While you are taking notes, you may revise or refine your thesis. After note-taking you should polish the thesis into one sentence to unite your findings. You may wish to add an organizational clue to the structure of your paper. For instance, your final thesis may read, "Computer access control systems demand constant upgrading due to the increasing amounts of sensitive computerized data, the expansion of networking systems, and the growing number of sophisticated, computer-competent criminals." This thesis contains not only an overall inference about your subject but also a clue to the sections of your report.

STRATEGY 3. RESEARCHING THE SOURCE MATERIAL

Objectives

In researching the source material you must accomplish three objectives:

1. ***Skim reading of the material.*** Skim read the material as you locate it in order to become familiar with the data on your subject. Look over the table of contents and index. Read the preface and introduction which may present an overview of the book's contents. Check the appendix and glossary for supplemental materials and definitions. Search for bibliographies that suggest more related sources for you to investigate.

2. ***Refinement of your thesis.*** Refine and amplify your thesis as you learn more about your subject.

3. ***Preparation of bibliography cards.*** Prepare a bibliography card for each source which contains information in support of your thesis so that you can quickly relocate the material for note-taking after your overview of materials.

Library

Familiarize yourself with your school, company, or public library. Locate the card catalogue, the reference work stacks, the indexes to periodical literature, the government publication indexes, the vertical file, the microfilm readers, and the photocopying equipment. Determine if the library catalogues its collection of books by the Dewey Decimal System or the Library of Congress System.

Dewey Decimal System. The Dewey Decimal System divides all books and journals and arranges them on the shelves by the following general subject categories:

000–099 General Works

100–199 Philosophy

200–299 Religion

300–399 Social Sciences

400–499 Language

500–599 Pure Sciences

600–699 Technology (Applied Sciences)

700–799 Fine Arts

800–899 Literature

900–999 History

Each of these general categories is, in turn, divided into ten smaller categories. For example the Technology (Applied Science) classification (600–699) is divided into the following:

600–609 Technology

610–619 Medical Sciences

620–629 Engineering

630–639 Agriculture

640–649 Home Economics

650–659 Business and Business Methods

670–679 Manufacturing (metal, textile, paper, etc.)

680–689 Miscellaneous Manufactures (hardware, furniture, etc.)

690–699 Building Construction

Library of Congress System. The Library of Congress System arranges the books on the shelves by classifying the books by 21 letters to designate the following major subject categories:

A General Works

B Philosophy and Religion

C History and Auxiliary Sciences

D Universal History and Topography

E/F American History

G Geography, Anthropology, Folklore, etc.

H Social Sciences

J Political Science

K Law

L Education

M Music

N Fine Arts

P Language and Literature

Q Science

R Medicine

S Agriculture

T Technology

U Military Service

V Naval Services

Z Library Science and Bibliography

Each of these classes is subclassed by a combination of letters for subtopics. For example T (Technology) contains 16 subclasses:

TA General engineering, including general civil engineering

TC Hydraulic engineering

TD Sanitary and municipal engineering

TE Highway engineering

TF Railroad engineering

TG Bridge engineering

TH Building construction

TJ Mechanical engineering

TK Electrical engineering, Nuclear engineering

TL Motor vehicles, Aeronautics, Astronautics

TN Mining engineering, Mineral industries, Metallurgy

TP Chemical technology

TR Photography

TS Manufactures

TT Handicrafts, Arts and crafts

TX Home economics

A number notation is also assigned to each book.

Card Catalogue. Begin your research by searching the card catalogue for reference and general books on your subject. The catalogue will contain three separate alphabetized cards for every book, filmstrip, phonograph album, and tape in the library. A separate card is filed under the author's name, title of the work, and subject heading. Thus, a book by Donn B. Parker titled *Fighting Computer Crime* will be catalogued under *P* for *Parker, Donn B.*, under *F* for *Fighting Computer Crime*, and under *C* for *Computer Crimes—Prevention.* Figure 12.1 shows a typical author card with explanations of the data:

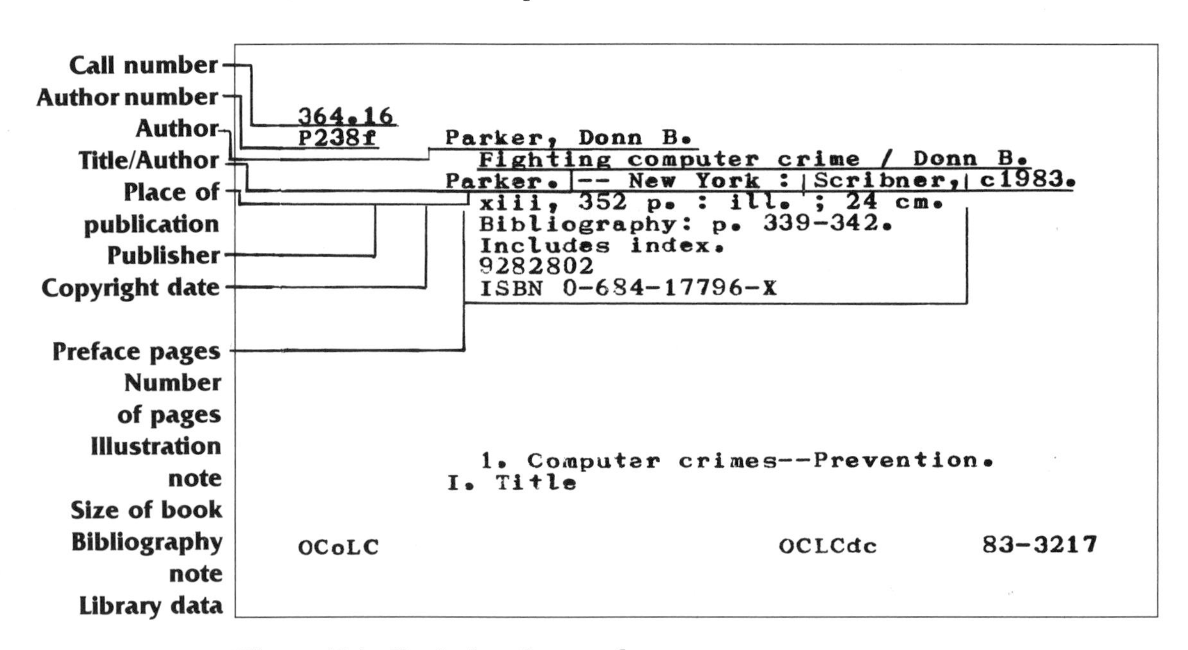

Figure 12.1 Typical author card

Figure 12.2 is the title card for the same book.

If you do not know the authors or titles of books on your subject, you may locate books by looking up *subject headings.* The subject heading "Computers," will be further divided into subcategories, such as "Com-

Fighting computer crime

364.16
P238f

Parker, Donn B.
Fighting computer crime / Donn B. Parker. -- New York : Scribner, c1983.
xiii, 352 p. : ill. ; 24 cm.
Bibliography: p. 339-342.
Includes index.
9282802
ISBN 0-684-17796-X

1. Computer crimes--Prevention.
I. Title

OCoLC OCLCdc 83-3217

Figure 12.2 Typical title card

puter Crimes—Prevention" and "Computers—Access Control." Figure 12.3 shows the subject card for the same book:

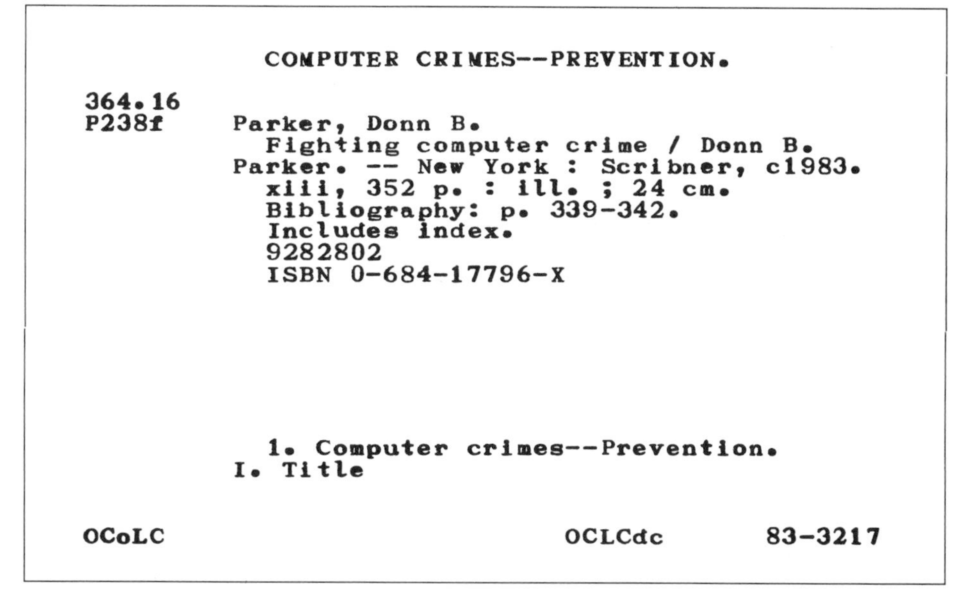

COMPUTER CRIMES--PREVENTION.

364.16
P238f

Parker, Donn B.
Fighting computer crime / Donn B. Parker. -- New York : Scribner, c1983.
xiii, 352 p. : ill. ; 24 cm.
Bibliography: p. 339-342.
Includes index.
9282802
ISBN 0-684-17796-X

1. Computer crimes--Prevention.
I. Title

OCoLC OCLCdc 83-3217

Figure 12.3 Typical subject card

Reference Books

Reference books—general encyclopedias, technical encyclopedias, almanacs, handbooks, dictionaries, histories, and biographies—are listed in the card catalogue. These are good sources to skim read first because the material is general and often contains bibliographies which will lead you to more specific sources. There are reference books for every discipline. Following is a partial list of technical reference books:

Encyclopedias

Encyclopaedia Americana

Encyclopaedia Britannica

Collier's Encyclopedia

Encyclopedia of Careers and Vocational Guidance

Encyclopedia of Computer Science

Encyclopedia of Food Technology and Food Service Series

Encyclopedia of Management

Encyclopedia of Marine Resources

Encyclopedia of Materials Handling

Encyclopedia of Modern Architecture

Encyclopedia and Dictionary of Medicine, Nursing, and Allied Health

Encyclopedia of Photography

Encyclopedia of Textiles

Encyclopedia of Urban Planning

McGraw-Hill Encyclopedia of Energy

McGraw-Hill Encyclopedia of Environmental Science

McGraw-Hill Encyclopedia of Food, Agriculture, and Nutrition

Dictionaries

Funk & Wagnall's New Standard Dictionary of the English Language

Oxford English Dictionary

Random House Dictionary of the English Language

Webster's Third New International Dictionary of the English Language

American Heritage Dictionary

Random House College Dictionary

Webster's New Collegiate Dictionary

Dictionary of Architecture and Construction

Dictionary of Business and Finance

Dictionary of Nutrition and Food Technology

Dictionary of Practical Law

Duncan's Dictionary for Nurses

Funk & Wagnall's Dictionary of Data Processing Terms

McGraw-Hill Dictionary of Scientific and Technical Terms

Paramedical Dictionary

Stedman's Medical Dictionary

Handbooks, Manuals, and Almanacs

Book of Facts

Building Construction Handbook

CRC Handbook of Marine Sciences

Fire Protection Handbook

Nurse's Almanac

U.S. Government Manual

The World Almanac

Indexes to Essays Within Books

Following your search of general and reference books on your subject, you may want to refer to one or more indexes of essays on your subject which are published within books of collected essays. Your search of the card catalogue subject cards may not turn up these essays unless the entire collection of essays is on one subject.

The *Essay and General Literature Index* catalogues essays by author and subject. It directs you to material within books and collections of a biographical and/or critical nature. Figure 12.4 shows a section of a page from the Essay and General Literature Index:

ESSAY AND GENERAL LITERATURE INDEX, 1975-1979 311

Composition (Music)—*Continued*
Sessions, R. Problems and issues facing the composer today. *In* Sessions, R. Roger Sessions on music p71-87
Sessions, R. Song and pattern in music today. *In* Sessions, R. Roger Sessions on music p53-70
Composition (Photography)
Borcoman, J. W. Notes on the early use of combination printing. *In* One hundred years of photographic history p15-18
Composition (Rhetoric) See Rhetoric
Compostela, Spain. See Santiago de Compostela, Spain
Comprehension
Bauman, Z. Consensus and truth. *In* Bauman, Z. Hermeneutics and social science p225-46
Bauman, Z. The rise of hermeneutics. *In* Bauman, Z. Hermeneutics and social science p23-47
Bauman, Z. Understanding as expansion of the form of life. *In* Bauman, Z. Hermeneutics and social science p194-224
Bauman, Z. Understanding as the work of history: Karl Mannheim. *In* Bauman, Z. Hermeneutics and social science p89-110
Bauman, Z. Understanding as the work of history: Karl Marx. *In* Bauman, Z. Hermeneutics and social science p48-68
Bauman, Z. Understanding as the work of history: Max Weber. *In* Bauman, Z. Hermeneutics and social science p69-88
Bauman, Z. Understanding as the work of life: From Schutz to ethnomethodology. *In* Bauman, Z. Hermeneutics and social science p172-93

Compulsory non-military service. See Service, Compulsory non-military
Computer composition. See Computer music
Computer music
Bamberger, J. In search of a tune. *In* Perkins, D. and Leondar, B. eds. The arts and cognition p284-319
Computer reliabilty. See Computers—Reliability
Computer translating. See Machine translating
Computers
Le Roy Ladurie, E. The historian and the computer. *In* Le Roy Ladurie, E. The territory of the historian p3-6
See also Electronic data processing
Reliability
Thomas, L. To err is human. *In* Thomas, L. The medusa and the snail p36-40
Computers and civilization
Lowi, T. J. The information revolution, politics, and the prospects for an open society. *In* Galnoor, I. ed. Government secrecy in democracies p40-61
Computers in literature
Rhodes, C. H. Tyranny by computer: automated data processing and oppressive government in science fiction. *In* Clareson, T. D. ed. Many futures, many worlds p66-93
Computing machines. See Calculating-machines
Comstock, W. Richard
On seeing with the eye of the native European. *In* Seeing with a native eye p58-78

Subject head
Author, article
Book author
Book title
Pages of essay
See also note
Subhead
Second entry

Figure 12.4 ***Sample Essay and General Literature Index page section (Copyright 1975, 1976, 1977, 1978, 1979 and 1980 by the H. W. Wilson Company. Materials reproduced by permission of the publisher.)***

At the back of each index volume is a "List of Books Indexed" which contains an alphabetized list by author of the books in which the essays appear. The call numbers of those books held by the library are often penciled in by librarians. If the call number is not penciled in, you may locate the book by checking to see if it is listed in the card catalogue.

Other indexes to books and collections are the *Bibliographic Index* and the *Biography Index: A Quarterly Index to Biographic Material in Books and Magazines.*

Abstracts

A library abstract is an index which, in addition to listing author's names, publishers, titles, and publication data, includes a brief summary of the content and scope of the book, article, or pamphlet. By skim reading the summaries, you can determine if the work is relevant to your subject. Do not take notes from an abstract; always cite material from the original work. Some technical abstracts include

Abstract on Criminology and Penology

Air Pollution Abstracts

Biological Abstracts

Criminal Justice Abstracts

Index to Publications of the United States Congress

Nursing Research

Oceanic Abstracts

Solar Energy Update

Figure 12.5 shows a section of a page from *Psychological Abstracts:*

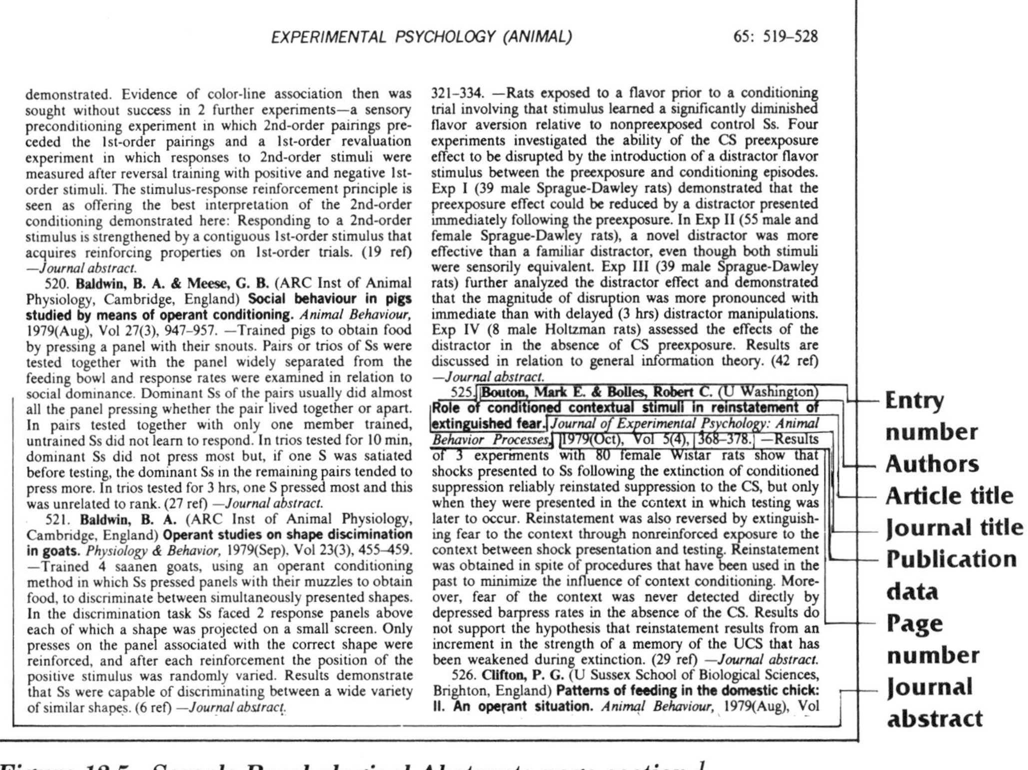

EXPERIMENTAL PSYCHOLOGY (ANIMAL) 65: 519–528

demonstrated. Evidence of color-line association then was sought without success in 2 further experiments—a sensory preconditioning experiment in which 2nd-order pairings preceded the 1st-order pairings and a 1st-order revaluation experiment in which responses to 2nd-order stimuli were measured after reversal training with positive and negative 1st-order stimuli. The stimulus-response reinforcement principle is seen as offering the best interpretation of the 2nd-order conditioning demonstrated here: Responding to a 2nd-order stimulus is strengthened by a contiguous 1st-order stimulus that acquires reinforcing properties on 1st-order trials. (19 ref) *—Journal abstract.*

520. **Baldwin, B. A. & Meese, G. B.** (ARC Inst of Animal Physiology, Cambridge, England) **Social behaviour in pigs studied by means of operant conditioning.** *Animal Behaviour,* 1979(Aug), Vol 27(3), 947–957. —Trained pigs to obtain food by pressing a panel with their snouts. Pairs or trios of Ss were tested together with the panel widely separated from the feeding bowl and response rates were examined in relation to social dominance. Dominant Ss of the pairs usually did almost all the panel pressing whether the pair lived together or apart. In pairs tested together with only one member trained, untrained Ss did not learn to respond. In trios tested for 10 min, dominant Ss did not press most but, if one S was satiated before testing, the dominant Ss in the remaining pairs tended to press more. In trios tested for 3 hrs, one S pressed most and this was unrelated to rank. (27 ref) *—Journal abstract.*

521. **Baldwin, B. A.** (ARC Inst of Animal Physiology, Cambridge, England) **Operant studies on shape discimination in goats.** *Physiology & Behavior,* 1979(Sep). Vol 23(3), 455–459. —Trained 4 saanen goats, using an operant conditioning method in which Ss pressed panels with their muzzles to obtain food, to discriminate between simultaneously presented shapes. In the discrimination task Ss faced 2 response panels above each of which a shape was projected on a small screen. Only presses on the panel associated with the correct shape were reinforced, and after each reinforcement the position of the positive stimulus was randomly varied. Results demonstrate that Ss were capable of discriminating between a wide variety of similar shapes. (6 ref) *—Journal abstract.*

321–334. —Rats exposed to a flavor prior to a conditioning trial involving that stimulus learned a significantly diminished flavor aversion relative to nonpreexposed control Ss. Four experiments investigated the ability of the CS preexposure effect to be disrupted by the introduction of a distractor flavor stimulus between the preexposure and conditioning episodes. Exp I (39 male Sprague-Dawley rats) demonstrated that the preexposure effect could be reduced by a distractor presented immediately following the preexposure. In Exp II (55 male and female Sprague-Dawley rats), a novel distractor was more effective than a familiar distractor, even though both stimuli were sensorily equivalent. Exp III (39 male Sprague-Dawley rats) further analyzed the distractor effect and demonstrated that the magnitude of disruption was more pronounced with immediate than with delayed (3 hrs) distractor manipulations. Exp IV (8 male Holtzman rats) assessed the effects of the distractor in the absence of CS preexposure. Results are discussed in relation to general information theory. (42 ref) *—Journal abstract.*

525. **Bouton, Mark E. & Bolles, Robert C.** (U Washington) **Role of conditioned contextual stimuli in reinstatement of extinguished fear.** *Journal of Experimental Psychology: Animal Behavior Processes,* 1979(Oct), Vol 5(4), 368–378. —Results of 3 experiments with 80 female Wistar rats show that shocks presented to Ss following the extinction of conditioned suppression reliably reinstated suppression to the CS, but only when they were presented in the context in which testing was later to occur. Reinstatement was also reversed by extinguishing fear to the context through nonreinforced exposure to the context between shock presentation and testing. Reinstatement was obtained in spite of procedures that have been used in the past to minimize the influence of context conditioning. Moreover, fear of the context was never detected directly by depressed barpress rates in the absence of the CS. Results do not support the hypothesis that reinstatement results from an increment in the strength of a memory of the UCS that has been weakened during extinction. (29 ref) *—Journal abstract.*

526. **Clifton, P. G.** (U Sussex School of Biological Sciences, Brighton, England) **Patterns of feeding in the domestic chick: II. An operant situation.** *Animal Behaviour,* 1979(Aug), Vol

Figure 12.5 Sample Psychological Abstracts page section [1]

[1]These citations are reprinted with permission of the American Psychological Association, publisher of *Psychological Abstracts* and the PscyINFO Database (Copyright © by the American Psychological Association), and may not be reproduced without its prior permission.

Indexes to Periodical Literature

In addition to books, libraries also subscribe to and hold periodical literature: magazines, journals, and newspapers. The articles in these various periodical indexes are listed by subject, author, and sometimes titles (books and films). Scan the library's periodical holdings list to determine which magazines, journals, and newspapers are bought by your library. Some of the materials will be microfilm copies of the periodicals. Periodical literature can not generally be checked out of a library, so be prepared to photocopy those articles from which you want to take notes.

Magazines and Journals

The Reader's Guide to Periodical Literature

Social Sciences Index (formerly the *International Index* and the *Social Sciences and Humanities Index*)

Accountant's Index

Applied Science and Technology Index

Biological and Agricultural Index

Business Periodicals Index

Criminal Justice Periodical Index

Cumulative Index to Nursing and Allied Health Literature

Education Index

Energy Index

Environmental Index

Hospital Literature Index

Index to U.S. Government Periodicals

Index to Legal Periodicals

Index Medicus

International Nursing Index

Microcomputer Index

The *Reader's Guide to Periodical Literature* is the most comprehensive. It indexes articles from over 150 popular magazines and journals and is published every few weeks, helping you to locate very recent articles. Figure 12.6 shows a section of a page from *The Reader's Guide:*

AUGUST 1983 117

Compensation (Law)—See also—*cont.*
Insurance, Workers' compensation
Compensatory education
Reagan sends Congress a voucher plan for compensatory education. *Phi Delta Kappan* 64:668-9 My '83
Competency tests *See* Educational tests and measurements
Competency tests for teachers *See* Teachers—Examinations
Competition
See also
Business intelligence
Trade marks and trade names
Competition (Biology)
Female moorhens compete for small fat males. M. Petrie. bibl f il *Science* 220:413-15 Ap 22 '83
Finches show competition in ecology [interspecific competition in Galápagos finches; work of P. R. Grant and others] R. Lewin. il map *Science* 219:1411-12 Mr 25 '83
Competition (Psychology)
See also
Sports—Psychological aspects
Good guys make great enemies. A. Brandt. il *Esquire* 99:20+ Ap '83
Competitions
See also
Advertising—Prize contests
Beauty contests
See also subhead Competitions under various subjects
All-American Girl. il *Teen* 27:82-5 Je '83
Compiègne Château *See* Castles—France
Complaints
Who do I call to complain? [consumers] il *Consum Res Mag* 66:14-17+ Je '83
Complexion *See* Skin
Component television *See* Television receivers—Components
Composers, American
See also
Argento, Dominick
Becker, John J., 1886-1961
Schwantner, Joseph C.
Composers, Canadian
See also
Schafer, R. Murray
Composers, German
See also
Trojahn, Manfred
Composers, Italian
See also
Busoni, Ferruccio, 1866-1924
Nono, Luigi
Composite materials
Europe's composite work stresses production gains. M. Feazel. il *Aviat Week Space Technol* 118:82-3+ Ap 18 '83

Taking a peek at computer camps. il *Seventeen* 42:98 Ap '83
Computer crimes
See also
Computer programming—Unauthorized use
Computer games *See* Video games
Computer graphics
Computer-aided logic design. C. Rubenstein. il *Comput Electron* 21:67+ My '83
Computer art aids surgeons [facial reconstruction] E. Rosenthal. il *Sci Dig* 91:72-3 My '83
Design automation for integrated circuits. S. B. Newell and others. bibl f il *Science* 220:465-71 Ap 29 '83
Machine-made body joints [computer assisted design and manufacture] il *Sci Dig* 91:89 Mr '83
The personal challenge of CAD [computer-aided design] B. Milliken. il *Archit Rec* 171:41+ Mr '83
Computer industry
See also
American Telephone & Telegraph Co.
Apple Computer Inc.
Comdisco, Inc.
Computer service industries
Convergent Technologies
Data General Corp.
Datapoint Corp.
Digital Equipment Corp.
Honeywell Inc.
Intelligent Systems Corp.
International Business Machines Corp.
Nanes Finishing and Assembly Corporation
NCR Corp.
Osborne Computer Corp.
Prime Computer, Inc.
Tandy Corp.
Texas Instruments Incorporated
Trilogy Systems Corporation
Vermont Research Corp.
Computer boom town [Lowell, Mass.] S. Solomon. il *Sci Dig* 91:38+ Mr '83
Computers: a crash plan to foil Japan. il *U S News World Rep* 94:8 My 30 '83
The U.S. studies its options. C. Norman. *Science* 220:799 My 20 '83
Antitrust cases
IBM: "Sit up and take notice" [antitrust claim brought by the European Economic Community] R. Greene and J. Bamford. il *Forbes* 131:176 My 23 '83
Cooperation
See also
Microelectronics and Computer Technology Corporation
Employees
See Computer personnel
Marketing

Subject
See also guide to related subjects
Article's title
Author
Illustration notation
Periodical title plus Volume: pages and date

Figure 12.6 ***Sample Readers' Guide page section (Copyright 1983 by the H. W. Wilson Company and reproduced by permission of the publisher.)***

All of the specialized subject indexes are similar to *The Reader's Guide.*

Several indexes to the articles printed in newspapers are generally available in libraries.

Newspapers

The New York Times Index

Newspaper Index (Washington Post, The Chicago-Tribune, New Orleans Times-Picayune, and *The Los Angeles Times)*

Wall Street Journal Index

In addition, every newspaper indexes its own publication. These indexes are available at a newspaper office section, "the morgue." Companies also index their newsletters. Both newspaper offices and companies will assist you with their indexes.

The New York Times Index is the most comprehensive of the large indexes. It indexes stories and articles which have appeared in the *Times* since 1851 by subject and author. The index also summarizes articles and occasionally reprints photographs, maps, and other illustrations which accompanied the original article. Figure 12.7 shows a section of a page of *The New York Times Index:*

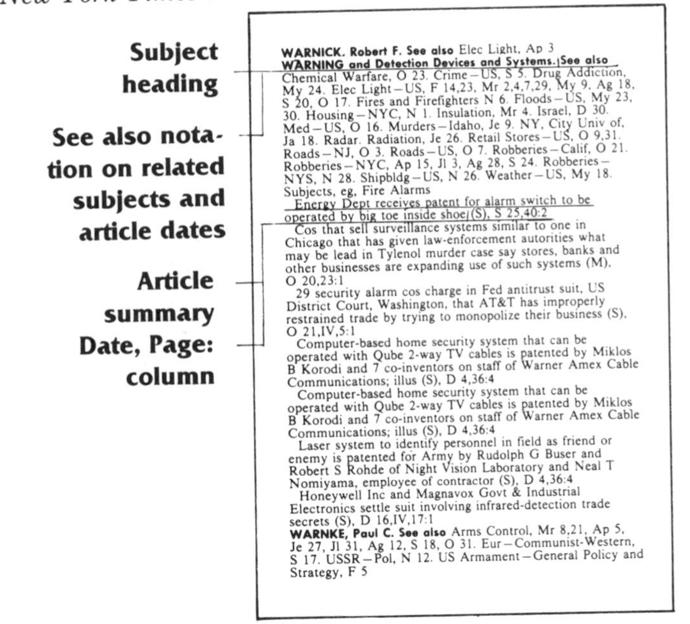

WARNICK, Robert F. See also Elec Light, Ap 3
WARNING and Detection Devices and Systems. See also Chemical Warfare, O 23. Crime—US, S 5. Drug Addiction, My 24. Elec Light—US, F 14,23, Mr 2,4,7,29, My 9, Ag 18, S 20, O 17. Fires and Firefighters N 6. Floods—US, My 23, 30. Housing—NYC, N 1. Insulation, Mr 4. Israel, D 30. Med—US, O 16. Murders—Idaho, Je 9. NY, City Univ of, Ja 18. Radar. Radiation, Je 26. Retail Stores—US, O 9,31. Roads—NJ, O 3. Roads—US, O 7. Robberies—Calif, O 21. Robberies—NYC, Ap 15, Jl 3, Ag 28, S 24. Robberies—NYS, N 28. Shipbldg—US, N 26. Weather—US, My 18. Subjects, eg, Fire Alarms

Energy Dept receives patent for alarm switch to be operated by big toe inside shoe (S), S 25,40:2

Cos that sell surveillance systems similar to one in Chicago that has given law-enforcement autorities what may be lead in Tylenol murder case say stores, banks and other businesses are expanding use of such systems (M), O 20,23:1

29 security alarm cos charge in Fed antitrust suit, US District Court, Washington, that AT&T has improperly restrained trade by trying to monopolize their business (S), O 21,IV,5:1

Computer-based home security system that can be operated with Qube 2-way TV cables is patented by Miklos B Korodi and 7 co-inventors on staff of Warner Amex Cable Communications; illus (S), D 4,36:4

Computer-based home security system that can be operated with Qube 2-way TV cables is patented by Miklos B Korodi and 7 co-inventors on staff of Warner Amex Cable Communications; illus (S), D 4,36:4

Laser system to identify personnel in field as friend or enemy is patented for Army by Rudolph G Buser and Robert S Rohde of Night Vision Laboratory and Neal T Nomiyama, employee of contractor (S), D 4,36:4

Honeywell Inc and Magnavox Govt & Industrial Electronics settle suit involving infrared-detection trade secrets (S), D 16,IV,17:1

WARNKE, Paul C. See also Arms Control, Mr 8,21, Ap 5, Je 27, Jl 31, Ag 12, S 18, O 31. Eur—Communist-Western, S 17. USSR—Pol, N 12. US Armament—General Policy and Strategy, F 5

Figure 12.7 Sample New York Times Index page section (Copyright 1983 by the New York Times Company. Reprinted by permission.)

Vertical File

Most libraries maintain a vertical file by subject of pamphlets, booklets, bulletins, and clippings on timely subjects. Ask the reference librarian to assist you in locating the file and the material contained within it.

STRATEGY 4. PREPARING A WORKING BIBLIOGRAPHY

Bibliography Cards

As you skim-read your source material, you should prepare a separate, annotated bibliography card on 3- by 5-inch cards for each source that appears to be useful. A bibliography is a compilation of background works or reading lists. An annotation is a note to yourself about the pages, chapters, special strengths, or other features of each source. As you review your material, you will frequently find bibliographies or reference lists at the end of books and articles you may want to investigate. Prepare a card for each new potential source in order to locate that material later. When you complete your research paper, you will organize the source cards you actually used into alphabetical order to type the final "Works Cited" list (see page 317).

Write the information on the card according to one of three styles: the new Modern Language Association (MLA) style, the Author/Date style, or the Number style.

New MLA Style for "Works Cited"

The Modern Language Association style is the most widely used system for documentation format; however, various disciplines frequently employ different systems of documentation. Recently, MLA developed a new, official style to simplify documentation procedures. Because a considerable amount of your research is bound to involve works documented according to the former system, you will want to familiarize yourself with the former MLA style as well as the new style. The former style is described on pages 348–53.

The following is a guide to the new MLA style for a "Works Cited" listing; Figure 12.8 shows a typical bibliography card for a book with one author:

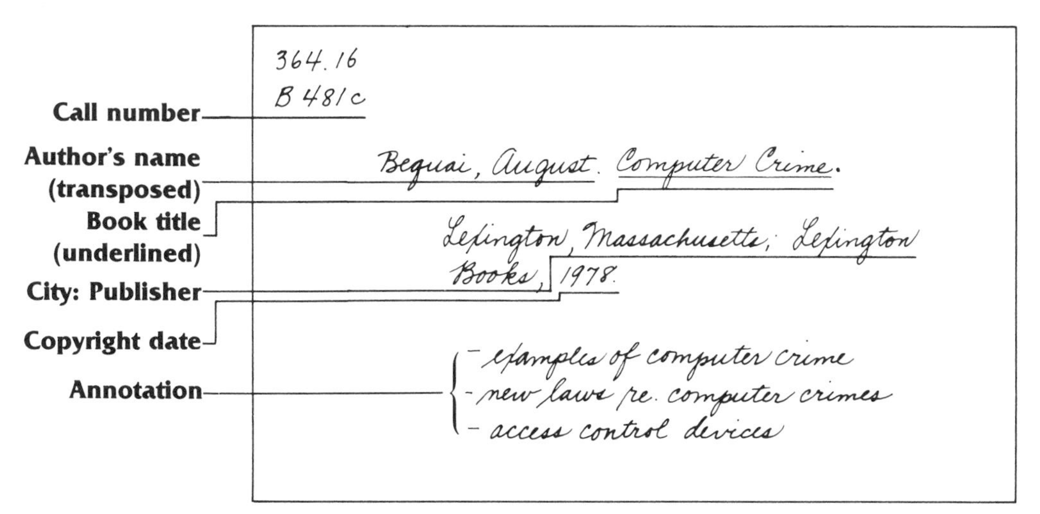

Figure 12.8 Typical bibliography card for a book with one author

Books

1. *Book with one author (Figure 12.8)*

2. *Book with two authors*

Hemphill, Charles F., and Robert D. Hemphill. Security Safeguards for the Computer. New York: Amocon, 1979.

3. *Book with three authors*

Hsiao, David K., Douglas S. Kerr, and Stuart E. Madnick. Computer Security. New York: Academic Press, 1979.

or

Hsiao, David K., and others. Computer Security. New York: Academic Press, 1979.

4. *Edited book*

Shank, Roger C. Computer Models of Thought and Language. Ed. by Roger C. Shank and Kenneth Mark Colby. San Francisco: W. H. Freeman, 1973.

5. *Book with corporate authorship*

Management Information Corporation. Computer Privacy. Cherry Hill, N.J.: Management Information Corporation, 1982.

6. *Essay in an edited collection*

Le Roy, Ladurie Emmanuel. "The Historian and the Computer." In Territory of the Historian. Ed. by Ladurie E. Le Roy. Chicago: University of Chicago Press, 1979, 16–36.

7. *Translated book*

Cardoza, Juan. Access Control of Computers. Trans. by Alan Jameson. New York: New American Library, 1981.

8. *Work of several editions or volumes*

Parker, Donn B. Crime by Computer. 2nd ed. New York: Scribners, 1976.

Anderson, Martin. Computer Networks. 2 vols. Englewood Cliffs. N.J.: Prentice-Hall, Inc., 1981.

Encyclopedias, Yearbooks, Dictionaries

9. *General encyclopedia article*

Smythe, Charles John. "Computers." Encyclopaedia Britannica. 1983 ed.

10. Specialized encyclopedia article

Parker, D. B. "Crime and Computer Security." Encyclopedia of Computer Science. Ed. by Anthony Ralston and Charles E. Meeks. New York: Mason/Charter Publishers, Inc., 1976.

11. Statistical Abstract

"Computer Services." Statistical Abstract of the United States, 82 (1979): 72.

12. Yearbook

"Another Wire-transfer Theft." Facts On File, 1979, 239.

Periodicals

13. Unsigned weekly magazine title

"Locking the Electronic File Cabinet." Business Week 18 Oct. 1982: 123–124.

14. Signed weekly magazine

Taflich, P. "Opening the Trapdoor Knapsack," Time 25 Oct. 1982: 88.

15. Monthly magazine

Lord, K. M. "Fingerprint Scanner: Security System With a Personal Touch." Popular Mechanics Oct. 1982: 128.

16. Journal with continuous pagination

Kolata, G. "Students Discover Computer Threat." Science, 2115 (1982): 1216–17.

17. Signed newspaper article

O'Neill, Robert. "Computer-based Home Security Systems Sales Soar." New York Times 15 Dec. 1983: sec. 2: 7.

18. Unsigned newspaper article

"Computer Security Concerns U.S. Military." Fort Lauderdale News/ Sun-Sentinel 25 Sept. 1983: sec. B: 4.

Miscellaneous Sources

19. Bulletin

Brown, Earl. "Computer Laboratory Security Measures." Gainesville, Fla., 1984. (Bulletin of the University of Florida, No. 72.)

20. ***Public document***

U.S. Congress. House. Committee on Interstate and Foreign Commerce. Federal Cigarette Labeling and Advertising Act. House Report 449 to accompany H.R. 3014, 89th Congress, 1st Session. 1965.

21. ***Pamphlet***

Radio Shack, A Division of Tandy Corporation. TRS-80 Model II Micro-Computer System. U.S.A. 1981. (Pamphlet of Radio Shack, A Division of Tandy Corporation.)

22. ***Letter***

Larsen, Robert H. Information in a letter to the author. R. H. Larsen and Associates, Fort Lauderdale, Fla., 11 Mar. 1984.

23. ***Interview***

Lessin, Arlen, President of Smart Card International, New York. Personal interview on microchip security. 12 Apr. 1984.

24. ***Photocopied source***

Baird, Jane. "Access Control Systems." Pompano Beach, Fla., 14 Mar. 1975. (Photocopied.)

25. ***Public address or lecture***

Davis, Ralph. "The Value of Grading by Computers." Gainesville, Fla., 12 May 1981. (Address presented at the University of Florida Staff and Program Development Seminar.)

26. ***Telecast or radio broadcast***

"Should We Get On with Computer Literacy?" The Firing Line. Washington, D.C.: PBS-TV, 16 Oct. 1983.

27. ***Tables or illustrations***

Buckley, Charles L. Accounting system flowchart. Introduction to Accounting. New York: Oxford University Press, 1980.

Final Format

When you type your final research report, include an alphabetized list of all of the works actually used in your study. This list should include only those works from which you quoted or paraphrased material. Title your list "Works Cited." A sample is shown in Figure 12.9.

The third entry indicates a second work by the above-named author (Cardoza, Juan). The underline is used to highlight that the entry is alphabetized by the titles of Cardoza's works.

WORKS CITED

Cardoza, Juan. Access Control of Computers. Trans. Alan Jameson. New York: New American Library, 1981.

______________. Computer Security. Englewood Cliffs, N.J.: Prentice-Hall, Inc., 1982.

"Computer Security Concerns U.S. Military." Fort Lauderdale News/Sun-Sentinel 24 Sept. 1983, sec. B : 7.

Hsiao, David K., and others. Computer Security. New York: Academic Press, 1979.

"Locking the Electronic File Cabinet." Business Week 18 Oct. 1982:123–126.

Parker, Donn B. Crime By Computer. 2nd ed. New York: Scribners, 1976.

Figure 12.9 Sample works cited listing

Author/Date Style

If you are writing a paper or report in the field of biological science, physical science, mathematics, or psychology, consult your instructor to determine the preferred style. Modification of style exists from field to field. In the *Author/Date system* you generally

- capitalize only the first word of book and article titles,
- or omit the periodical article title entirely,
- abbreviate the title of a periodical, and
- omit the underline of a periodical title, but
- underline the periodical volume number.

The following bibliography entries illustrate a typical book and periodical entry for several scientific fields:

Biology

Lewin, R. A. 1982. Symbiosis and parasitism; definitions and evaluations. Bio Science. 32: 254.

Mader, Sylvia S. 1976. Inquiry into life. W. C. Brown Co., Dubuque, Iowa. 740 p.

Chemistry

J. M. Widom, "Chemistry, an Introduction to General, Organic, and Biological Chemistry." W. H. Freeman, San Francisco, 1981.

R. Seeber and S. Stefani, Anal. Chem., 53, 1011–1016 (1981).

Geology

Thomas, W. L. Jr. (ed.) 1956. Man's role in changing the face of the earth: Chicago, University of Chicago Press, 437 p.

Brocker, W. S. 1970. Man's oxygen reserves. Science, v. 168, p. 1537–1538.

Mathematics

D. G. Crowdis, Precalculus mathematics, Glencoe Press, Beverly Hills, Calif., 1976.

S. MacLane, Mathematical models: a sketch for the philosophy of mathematics, Amer. Math. M. 88 (1981), 462–472.

Physics

L. A. Marschall, Am. J. Phys. 49, 557–561 (1981).

S. Gasiorowicz, The Structure of Matter: a Survey of Modern Physics (Addison-Wesley Publishing Company, Reading, Massachusetts, 1979).

Psychology

Bugelski, B. R., and Graziano, A. M. Handbook of practical psychology. Englewood Cliffs, N.J.: Prentice-Hall, Inc., 1980.

Szucho, J. J., and Kleinmuntz, B. Statistical versus clinical lie detection. American Psychologist, 1982, 36, 488–96.

Alphabetize your "List of References" at the end of your paper.

Number Style

The bibliography format of the *Number System* is similar to the Author/Date system. However, in the compiled "List of References," each entry is numbered. The list may be alphabetized and numbered consecutively, or the list may forego an alphabetical listing and be numbered and listed as they are referred to chronologically in the text.

STRATEGY 5. BRAINSTORMING THE WRITING STRATEGIES

Having located and reviewed the source materials and considered your preliminary thesis, you will proceed to brainstorm the writing strategies needed to prove your thesis.

Ask yourself these questions:

1. Will I need to define terms?
2. How will I classify and divide the subtopics of my subject?
3. What kinds of examples will support my thesis?
4. What comparisons and contrasts help to explain my material?
5. Are cause-and-effect factors an essential part of my evidence?
6. Will I need to describe or analyze a process in my paper?

Organize these strategies into a logical presentation.

Think ahead, too, to the type of graphics you will want to seek out or devise. With these writing and illustrating strategies in mind, you are ready to prepare an outline.

STRATEGY 6. PREPARING THE WORKING OUTLINE

Develop a preliminary or working outline of your materials. Consider the writing strategies which you have brainstormed. Your first draft may be sketchy, but it will serve as a guideline to your note-taking. Without an outline you are likely to take notes on materials irrelevant to your purpose or overlook an area which should be explored.

You should not be totally bound by your working outline. As you think over your subject and take notes, you will want to expand, rearrange, discard, and subordinate your outline topics. Your final research report should include a final outline.

Format

Select a traditional or decimal outline format:

I.
    A.
        1.
        2.
    B.
        1.
        2.
            a.
            b.
                (1)
                (2)
                    (a)
                    (b)

1.0
    1.1
        1.1.1.
        1.1.2.
    1.2
        1.2.1.
        1.2.2.
2.0
    2.1
        2.1.1.
        2.1.2.

Although the Roman-numeraled, traditional outline allows for five levels of subordination, you will seldom find it necessary to use subheads beyond two subordinates. The decimal outline format is popular for technical presentations.

Topic Outline

Although each entry in an outline may be a sentence, a topic outline is not only easier to develop in the preliminary stage but also more common in the final draft. Major topics name the divisions of your paper. Next, subtopics are subordinated and indented under each major topic, and subsequent developmental topics are subordinated beneath each subtopic.

Partial Preliminary Outline

1.0 Introduction (thesis)
  1.1 Computer definition
  1.2 Access control definition
    1.2.1 Access control examples
    1.2.2 Access control processes
2.0 Causes of computer security violations
  2.1 Cause 1
  2.2 Cause 2
  2.3 Cause 3
3.0 Effects of computer security violations
  3.1 Effect 1
  3.2 Effect 2
  3.3 Effect 3
4.0 Examples of computer crimes
  4.1 Industrial crimes
    4.1.1 Company A
    4.1.2 Company B . . . (etc.)

Keep working on your outline as you take notes.

Partial Final Outline

2.0 Causes of computer security violations
  2.1 Faulty access control devices
    2.1.1 Telephone networking
    2.1.2. Passwords
  2.2 Excessive computerized information
    2.2.1 Pentagon's 8000 computers
    2.2.2 Rand Corporation statistics
  2.3 Number of trained computer specialists
    2.3.1 Number of military contractor clearances
    2.3.2 Number of U.S.-trained computer operators
  2.4 Amateur experimenters
    2.4.1 Milwaukee amateur club incident
    2.4.2 Soviets in Applied Systems Analysis group

STRATEGY 7. TAKING NOTES

Keeping your thesis in mind at all times and using your working outline as a guide, you now begin to record the notes from each of your sources. Notes are written on 3 by 5-inch or 4 by 6-inch index cards. Use a consistent system to record your information:

1. Enter only one item of information on each card so that you can shuffle and rearrange the cards during all stages of research.
2. Write on only one side of each card. A note continued on the back may be overlooked. If an item of information is too lengthy for one card, use two or more and staple them together.
3. Record the source at the top left of the card. This may be an abbreviated form of the bibliography card, such as the author's last name, the book or article title, or both.
4. Write the note in the center of the card. This may be a summary, a paraphrase, or a quotation. Not all of your notes will be jotted down in complete sentences; some may contain statistics, definitions, phrases, or other fragmentary information to be rewritten carefully in the report draft.
5. Record the exact page number(s) of the source from which you took the note.
6. Label the card in the upper right-hand corner with a word or two which indicates the topic of the note. The topic notation may correspond with an entry in your outline. Figure 12.10 shows a sample note card:

Source — Yeager, Corporate Crime

Topic Notation — Equity Funding Corp. Fraud

Note — Equity Fund Corp. of America (1973 case) fraudulently created 64,000 fictitious insurance policies out of a total 97,000. Computer printouts indicated the company was the fastest growing financial institution in the world.

Page — p. 6-7

Figure 12.10 Sample note card

Summary and Paraphrase Notes

Most of your notes will consist of summaries or paraphrases, that is, condensations or restatements of the source material in your own words. Following is an excerpt as it appears in the original source material. Notice that the note (Figure 12.11) both summarizes the information and restates the message in the note taker's own language.

Original

For workers one of the most fearful aspects of the computer is its capacity of controlling machinery to perform tasks that were formerly done by human labor. The unemployment that results is called *technological unemployment*. Employees who are eliminated by automation are generally unskilled and semiskilled workers who find difficulty in seeking new employment. Because of the growing use of computerized checkouts in supermarkets, the Retail Clerks Union fears a loss of 25 to 30% of supermarket jobs. Perhaps many of these fears are unjustified. The telephone companies now employ more persons than they did before the use of computerized switching services.

Glos and others Technological Unemployment

Because computers can perform tasks previously done by people, technological unemployment results. Unskilled and semiskilled workers lose their jobs and cannot find other employment. Despite the fact that telephone companies hire more people due to computerized switching services, workers' unions fear that employment levels will fall due to computerized services

309

Figure 12.11 Summary and paraphrased note

Direct Quotation

Occasionally, you will want to quote the exact words of an author. Usually no more than 10 percent of the text should consist of direct quotation. Take notes of direct quotation only if the material conveys a highly original idea, opinion, or conclusion of the author which, if paraphrased, would lose its striking effect or distort the meaning.

Following is an original excerpt plus two sample notecards (Figures 12.12 and 12.13), one employing full sentence quotation and the other employing partial sentence quotation. Note the use of the ellipsis (. . .) to indicate omitted words and the brackets ([ ]) to indicate slight additions by the note taker.

Original

Because of the sea of dangers, past practice has been to isolate sensitive computers. That strategy proved itself this summer at the Los Alamos National Laboratory in New Mexico when young raiders from Milwaukee succeeded only in cracking an unclassified computer. There is no evidence that they actually obtained classified information or gained access to the top-secret computers, which remain unconnected to networks.

Broad, Computer Security...

Military Concerns

Computer crime researcher, William J. Broad, notes, "Because of the sea of dangers [to computer security], past practice has been to isolate sensitive computers."

40

Figure 12.12 Full sentence quotation note

Broad, Computer Security...

Military Concerns

Researcher William J. Broad views the possibility of computer break-ins as "a sea of dangers." He documents a computer crime at the Los Alamos Nat'l Lab. in which "young raiders... succeeded... in cracking an unclassified computer. [But]... they [did not] actually obtained classified information...."

40

Figure 12.13 Partial quotation note

Plagiarism

Notes must be recorded carefully to avoid plagiarism. Plagiarism is offering the words or ideas of another person as if they were your own. The published words, ideas, and conclusions of an author are protected by law and must be acknowledged whenever you use them as evidence. To avoid plagiarism, follow five rules:

1. Introduce all borrowed material by stating in the text the name of the authority from whom it was taken.
2. Enclose the exact words within quotation marks.
3. If you summarize or paraphrase material, make sure that the information is written in your own style and language.
4. Provide a parenthetical reference note for each borrowed item.
5. Provide a "works cited" entry for every source which is referenced in your text.

Following is an original excerpt followed by three note cards. The first two (Figures 12.14 and 12.15) are unacceptable plagiarisms, while the last (Figure 12.16) is an acceptable note.

Original

The explosive development of the computer has created change both inside and outside an organization. Inside the organization the computer has altered work assignments and provided management with new tools for decision making. Outside the organization the computer is greatly changing our way of life and will continue to exert a force for change. Some of the social concerns related to the computer are: generation of unnecessary information, organizational changes, fear of technological unemployment, invasion of privacy, security of data, and depersonalization.

Business: Its Nature... Social Impact of Computer

The explosive development of the computer has caused business and society to change. In business, the computer has changed work assignments and management decision making methods. In society the computer has raised concern about unnecessary info, organizational change, technological unemployment, privacy, security of data, and depersonalization.

473

Figure 12.14 Plagiarized note

Business: Its Nature... Social Impact of Computers

Computers are creating change in and out of business. Computers exert a force for change for methods of work assignments, decision making information storage, organizational structures, technological unemployment, privacy, data security, and depersonalization.

473

Figure 12.15 Less obvious, but plagiarized note

Business: Its Nature... Social Impact of Computers

Computers are revolutionizing society. Professors Glos, Steade, and Lowry point out that " the explosive development " of computers affects not only the businessman but also all of us. They list seven major concerns including changes in management methods, excessive information, employment layoffs, and privacy and security worries.

473

Figure 12.16 Acceptable note

In the acceptable note (1) the information is summarized, (2) the language is the note taker's, (3) the authors' phrase is quoted, and (4) the source is acknowledged in the text. In addition, a parenthetical reference to the source and a complete "Works Cited" entry will appear in the final draft of the report.

STRATEGY 8. WRITING THE ROUGH DRAFT

Using your outline as a structural guide and your note cards as the content guide, write a rough draft of your research report. Follow these guidelines:

1. Type or write your draft leaving space between each line in order to expand or revise your draft.
2. Write your introduction. Be sure your limited subject and final thesis are clearly stated. A clue to the overall organization of the paper is needed if it is not indicated by your thesis. Clearly define your key terms before beginning to present your evidence which supports your thesis.
3. Write the body paragraphs. Each should begin with a topic sentence, a sentence which clarifies the topic of the paragraph and states your opinion or generalization about it. Each paragraph must have unity (address itself to the same subtopics of your overall subject), coherence (logical progression of evidence), and transitions (words and phrases which connect the evidence, such as *for instance, in the first place, furthermore, finally*).
4. When paraphrasing opinions and conclusions or directly quoting any source material, be sure to include the author's name in the text. Check the five safeguards against a plagiarism charge to ensure that all of your evidence is documented.
5. Insert your internal reference notes in parentheses as you write and circle each in order not to overlook one when you prepare your final draft.
6. Write your conclusion. This may be one or several paragraphs which summarize your findings, present solutions, or recommend a particular action.
7. Edit all mechanics (spelling and punctuation) and style considerations (sentence construction, grammar, and usage).

STRATEGY 9. REFERENCING THE SOURCES

Your notecards contain the author, work, and page number of each summarized, paraphrased, or quoted note. A parenthetical reference note for each cited source must be included with your report.

New MLA Reference Style

Page references for cited sources must be included *in parentheses* in the text with the author's last name or a short title of the work—or both—added as necessary.

The following is a guide to the MLA style:

1. Reference identification when the author's name is mentioned in the text.

> E. M. Johnson suggests that the increase in computer networking poses the greatest security problem (101).

or

E. M. Johnson suggests that the increase in computer networking poses the greatest security problem (Newsweek 101).

2. Reference identification when the title of the work, but not the author's name, is mentioned in the text.

The notion is put forth in Computer Security that computer passwords pose another problem (Lloyd 72).

3. Reference identification when the author and work are not mentioned in the text.

Because the data may be raided by unscrupulous individuals, computers may not be the cure-all for streamlining manufacturing procedures (Foster, Computers Today 19–20).

4. Reference identification when two works by the same author are listed in "Works Cited."

Marilyn Lanshe argues that technology can overcome computer security problems ("Beating the Risk" 78).

5. Reference identification when part of a larger work or specific volume is indicated.

(Computers 2: 214–15)

6. Reference identification in the middle of a sentence.

Following several data thefts from Ace Electronics (Laird, Computer Crime 142), the industry moved toward . . .

7. Reference identification when the text makes it clear that consecutive material is from the same source.

H. Charles Howard suggests that the military is experiencing "gargantuan problems" due to computer password systems. He stresses the necessity for changing passwords frequently and deplores the problem of dissemination of new passwords (36–40).

8. Reference identification when the text includes a display quote.

Rick Dyer, engineer and businessman, promotes Halycon:

> Halycon is not a game. Halycon is about total involvement. These [home video programs] are "adventures." You start to forget it's not real. You make the choices and the decisions. [Halycon] is mind-stimulating, not wrist-stimulating like video games. (Chicago Tribune D17)

9. Reference identification for figures (after title).

Figure 14 Schematic symbol for a resistor. From Kline (89).

10. ***Reference identification for tables (below table with asterisk).***

Table 2* Computer Systems Comparisons

Brand	Customer Accounts	Cost ($)
Apple	1,000	5,000
TRS-80	1,000	4,000
Altos	12,000	20,000

*From Consumer Reports (97)

Content Notes

Content footnotes—or, more usually, endnotes—may be used for definitions, explanations, acknowledgments, or amplifications of the text. Because such notes are distracting to readers, you should consider carefully whether the comments should be integrated into the text. Use only those notes that strengthen the discussion.

If you include content notes, type raised Arabic numeral superscripts at the points in the text that refer to the notes. Type the notes on a separate page(s) following the last page of the text. Title the page "Notes." Type content notes in the following form:

> [1]Briefly, the Keynesian theory states that a modern capitalistic society is naturally unstable because of a built-in lag in total demand.

The first line of a note is indented and is preceded by the raised superscript number that corresponds to the text reference.

A content note may also include a reference citation:

> [2]In addition, Keynesian theory has influenced other economic policies, notably the Full Employment and Balanced Growth Act (Rust, *Business in Society* 700).

Complete documentation of a content note source reference must be included in your "Works Cited" list.

Author/Date Style

In the Author/Date system the documentation consists of an alphabetized "List of References" (pp. 318-19) and internal reference notes in parentheses. Although this system is similar to MLA style, page references are excluded.

Guidelines for the author/date system follow:

1. If your text does not include the author's name, insert both the author's name and date within parentheses.
2. If your text includes the author's name, insert the publication date only within parentheses.
3. For a source with two authors, employ both names.
4. For three authors, name them all in the first instance, but thereafter use the first author's name and *et al.*
5. For four or more authors, employ the first author's name plus *et al.*
6. Use small letters (a, b, c) to identify two or more works published in the same year by the same author. (The "List of References" must include the letter after the publication year.)
7. Specify additional information (volume number, two works by the same author on the same subject published in different years, and pages) if necessary.

Following are sample internal notes in a text:

Author in text; year only

Argyle contends that women are more receptive than men to nonverbal encoding and decoding (1967). In a study using all male subjects, it was found that men who were good at displaying their emotions nonverbally

two authors

tended to be poor at understanding other nonverbal expressions (Lanzetta and Kleck, 1970). Women are more able to interpret the nonverbal messages

three works, one which is by four or more authors

of other people (Argyle, 1967; Buch et al., 1972a; Faltico, 1969). For instance, smiling can release aggressive tension or soften the impact of hostile words

two works by same author in same year

(Mehrabian, 1971a). The meaning of a smile depends very much upon the context in which it is used since it has so many possible meanings (Mehrabian, 1971b). Mothers' smiles have little relationship to the verbal messages they

three authors; 3rd volume of work

give their children (Bugenthal, Love, and Gianetto, 1971, III).

Number System

Using the numbered "List of References" required for the Number System, insert the appropriate number within your text in parentheses. The following variations occur:

1. If the sentence construction names the author, insert the number immediately after the authority's name within parentheses.

Example

J. S. L. Gilmore (1) gives an exactly analogous circular definition. Simpson (2) also emphasizes the subjective element in species classifications.

2. If the sentence construction does not require the authority's name, either

a. insert both author and number within parentheses:

Example

But everyone (Gilmore, 1; Huxley, 3; Simpson, 2) agrees that the empirical material places large constraints upon the systematist's inclinations.

b. insert both author and number within parentheses and enclose the number within brackets:

Example

There is remarkable agreement among competent workers, at least in the fairly well-worked animal groups. One first examines the material, and then one exercises "flair" (Huxley [3]).

c. insert the number only, enclosing it within parentheses or brackets.

Example

It is known [1] that every ontogeny is a developmental history that begins either with the fusion of two cells or with the cleavage of an unfertilized female gamete.

STRATEGY 10. PREPARING THE FINAL DRAFT

Type your final draft and submit it in the proper order and fastened together in a binder. The research report generally consists of the following:

1. ***Title page***—a descriptive title, your name, name of the course or company, the name and title of the person to whom the report is submitted, and the date.

2. ***Outline***—a correctly punctuated, capitalized, and grammatically constructed format with the page numbers in lower-case Roman numerals.

3. ***Text of the report***—a manuscript with the title repeated on the first page and numbered sequentially in Arabic numerals.

4. Internal reference notes—appropriate parenthesized author, work, and page references within the text.

5. Content notes—consecutively numbered notes placed at the bottom of the appropriate page but more preferably on a separate page(s) entitled "Notes" or "Endnotes" following the text.

6. List of references—a separately titled page entitled "Works Cited" containing the alphabetized listing of source material used in the preparation of the manuscript.

Figure 12.17 shows a student research report which demonstrates title page, outline, body, and MLA documentation techniques:

THE TECHNICAL WRITING CONSULTANT:

AN ANALYSIS OF A CAREER

Submitted to

the Department of English

Broward Community College

In partial fulfillment

Of the Course English 2210–02

Gordon R. Blaise
April 15, 1983

Figure 12.17 Sample research and documented report (Courtesy of student Gordon R. Blaise)

Title is capitalized.

OUTLINE

Thesis statement

Thesis: An extreme few choose to pursue a career in the highly diverse field of technical writing consultation, because this branch of specialization is the most intense, demanding, and well-paid branch of professional writing.

Topic outline

Entries are parallel.

I. Introduction

II. Background of technical writing

A. Definition

B. Specialties

C. Origin

D. Importance

Outline shows three-stage development.

III. Technical writing consultants

A. Operations

1. Home

2. Agency

B. Advantages

1. Income

2. Variety

Lower-case Roman numeral

i

Figure 12.17 Continued

3. Contact ii
4. Independence

C. Types of writing

1. Instructions
2. Service manuals
3. Graphics
4. Editing
5. Binding
6. Production
7. Proposals
8. Corporate reports

IV. Required education

A. Degree
B. Courses
C. Colleges and universities

V. Salaries

A. Technical writers
B. Technical writing consultants

Figure 12.17 Continued

iii

VI. Employment

A. Numbers

B. Future

C. Places

VII. Conclusion

Figure 12.17 Continued

THE TECHNICAL WRITING CONSULTANT:

AN ANALYSIS OF A CAREER

Well-known statement. Needs no reference note.

It has been said that diversity is an attribute that ranks second to none. This is especially true of the professional technical writer because he is often faced with an assignment that requires him to write, edit, compile, and publish on a subject about which he has little or no knowledge. This dilemma often requires the writer to be an expert in the gathering and compilation of specialized research as well as a master of grammar and style. It is for this reason that many professional technical writers and editors choose to specialize in a specific area of expertise, such as data processing documentation, advertising, instructional writing, or technical editing. An extreme few, however, choose to pursue a career in the highly diverse field of technical writing consultation because this branch of specialization is the most intense, demanding, and well-paid branch of professional writing.

Introduction introduces the subject and leads reader to narrowed subject.

Thesis narrows the subject and states opinion.

Part II.A of outline

Technical writing itself can be formally defined as a specialized branch of professional writing that relates solely to the worlds of business, science and industry. It is the professional technical writer who prepares most of the "how to" publications, repair manuals, and procedures manuals necessary in our technological world.

Title page is numbered at bottom.

1

Figure 12.17 Continued

Consecutive pages are numbered here.

2

The term itself, technical writing, is a union of two words, technical and writing. The word technical is a direct descendant of the Greek word tecnikos, meaning "art or skill." Writing, a typical English word, has many meanings. In this application, however, the best meaning that can be applied is "the composition of words." The word comes to Modern English as a variation of the Old English word writan, meaning "to scratch or to draw." It is a common belief in modern historical linguistics that the origin of this word is derived from the Greek word rhine, which means "file of rasp" (Webster's 1688).

Reference note in parentheses

Paraphrased research data

Part II.B

Part II.C

Technical writing, as a formal professional position, is often called "professional writing." It is also quite often referred to by the names of its various specialties, such as "documentation writing" in the field of data processing or "editing" in the field of technical editing (Pearsall and Cunningham 1).

The origin of technical writing dates back several thousand years to the ancient scribes whose primary job was to write in the language of the people the "technical information" given them by the priests, teachers of the temples, and the government. As the centuries passed, the work of these scribes became more and more sophisticated, as did technology and industry. It was, however, not until the emergence of the Industrial

Figure 12.17 Continued

3

Revolution that the need for a more precise specialized form of written communication was needed. In our modern technological world technical writing is a specialty that offers many opportunities. Advertising, script writing, documentation writing, science writing, teaching, and technical journalism are only a few of the many specialized branches of this very open and diversified profession (Encyclopaedia Britannica 23: 819).

Reference note includes volume and page.

Part II.D

Technical writing is not journalism as is often thought by the general public. It is, however, similar to journalism in that it is a form of written communication that often involves a great deal of research and investigation. Like journalism, technical writing can be a very complicated process, the end result being a clear and concise edition on the specified subject. (Van-Alstyne) It does not, however, promote opinion or speculation on the intended subject as do certain kinds of journalism. According to Thomas E. Pearsall and Donald H. Cunningham in their book, How to Write for the World of Work, "It [technical writing] provides the records that industry...needs to function year by year" (1).

Direct quotation with brackets for writer's additions and ellipsis to indicate omission.

Part III.A

Question used for transition.

Who, then, is the technical writing consultant? He or she is a free-lance technical writer. The consultant may work from one's own home, write for an agency, or even manage one's own agency. Because industry and research-and-development organizations tend to hold down their own staffs, such

Figure 12.17 Continued

4

Reference note refers to interview

organizations often turn to consultants for writing, editing, illustrating, and printing services (VanAlstyne). The advantages of consulting work are, according to Emerson Clarke and Vernon Root in A Definitive Study of Your Future in Technical and Science Writing, a higher average income than the industrial technical writer, a greater variety of work, the opportunity for wider contact with more people, and a degree of independence (32).

Part III.B

The types of writing the consultant will undertake will range from sets of instructions for users of small appliances to highly detailed service manuals for technicians. Figure 1 shows a simple set of instructions for a thermal glass:

Part III.C

Introduction to graphic; note colon.

Graphic inserted as figure

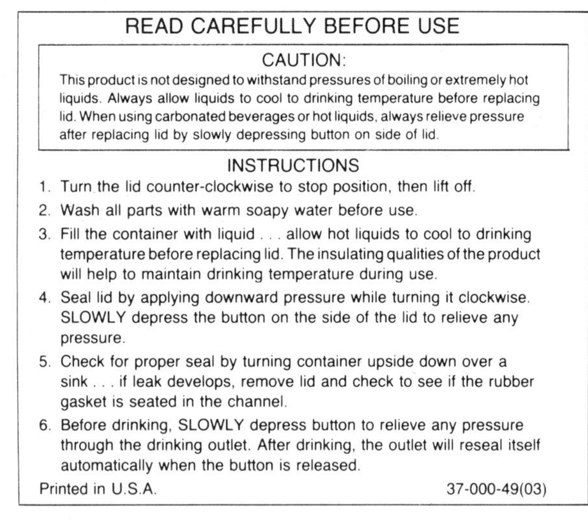

READ CAREFULLY BEFORE USE

CAUTION:
This product is not designed to withstand pressures of boiling or extremely hot liquids. Always allow liquids to cool to drinking temperature before replacing lid. When using carbonated beverages or hot liquids, always relieve pressure after replacing lid by slowly depressing button on side of lid.

INSTRUCTIONS

1. Turn the lid counter-clockwise to stop position, then lift off.
2. Wash all parts with warm soapy water before use.
3. Fill the container with liquid . . . allow hot liquids to cool to drinking temperature before replacing lid. The insulating qualities of the product will help to maintain drinking temperature during use.
4. Seal lid by applying downward pressure while turning it clockwise. SLOWLY depress the button on the side of the lid to relieve any pressure.
5. Check for proper seal by turning container upside down over a sink . . . if leak develops, remove lid and check to see if the rubber gasket is seated in the channel.
6. Before drinking, SLOWLY depress button to relieve any pressure through the drinking outlet. After drinking, the outlet will reseal itself automatically when the button is released.

Printed in U.S.A. 37-000-49(03)

Figure 1 A set of instructions for a product. From Thermo Serv Company.

Note reference note.

Figure 12.17 Continued

5

The independent consultant must take into account all of the many forms of professional writing, including graphics, editing, and often binding and production. It is these skills which permit the consultant to hire out to other businesses and industry. Many times, the technical writing consultant will form a partnership with other writers and offer a larger variety of services, such as proposal writing, corporate reports, and production (Publishers Weekly 18).

Part IV of outline

It should not come as a surprise that the individual pursuing a career in this field must first obtain a sound educational background. According to authors J.R. Gould and Wayne A. Losano, "Most technical writers are graduates of four-year programs and have earned bachelor's degrees" (Opportunities 29). Thomas E. Pearsall, technical writing educator, suggests that in addition to seventy quarter hours in a communication core, the student should complete four hours of math, four hours of computer science, eighteen hours of physical and biological science, and twenty hours of technical electives in the areas where one plans to work: computer science, biology, history of science, statistics, and the like (ADE Bulletin 17).

Direct quotation

Reference note shortens title

The number of universities offering technical communication degrees is growing. Following is a list of universities currently (1982) offering such degrees:

Figure 12.17 Continued

Informal table

6

Boston University

Bowling Green State University

Carnegie-Mellon University

Clarkson College

Colorado State University

Florida Institute of Technology

Los Angeles Trade-Technical College

Metropolitan State College

Miami University (Ohio)

Michigan Technological University

University of Michigan, Ann Arbor

University of Minnesota

New Mexico State University

North Carolina State University

Oklahoma State University

Oklahoma State University Technical Institute

Oregon State University

Pittsburg State University

Polytechnic Institute of New York

Renssalaer Polytechnic Institute

Figure 12.17 Continued

7

Reference note

Rock Valley College

South Dakota State University

University of South Dakota at Springfield

University of Wisconsin (17)

Many other college and universities offer certificates in technical report writing programs (VanAlstyne).

Part V of outline

The pay scale of the technical writer can best be described as being as diverse as is the field itself. For example, the average salary range for a technical writer in the employ of a large corporation ranges from $12.00 per hour to $30.00 per hour or $600 to $1200 per week, depending on education and experiences (Schemehauer 855). On the other hand, the earnings of the technical writing consultant are totally dependent upon the number of and the type of contracts he is able to obtain. Table 1 shows the average fees charged by consultants:

Table graphic

Table 1 Average Rate Scale. 1983*

Type of Contract	Average Fees ($)
Advertising Copywriting	10.00–36.00/hr.
Annual Reports	1500.00–7500.00
Audio Cassette Scripts	75.00–100.00/min
Business	200.00/day
Editing	200.00/day + exps.

Note table reference note

*From Schemehauer (855).

Figure 12.17 Continued

Part VI of outline

Reference note shortens corporate authorship

8

The number of employed technical writers has grown considerably over the past three years. In 1981 there were a total of 267,000 technical writers employed in a job market that demanded 288,000. This is in contrast to the 1983 statistics that showed a total market demand of 293,000 of which 292,000 jobs were actually filled (USDC 236).

The future of technical writing most probably lies in the field of consultation. Although the need for good technical writers grows with the expansion of science and industry, most of the world is suffering from a very serious recession. It is this unfortunate situation that often places the technical writer in a position of being a "luxury commodity" that many businesses cannot afford. It is this dilemma that often forces many such businesses to utilize the services of a consultant as the need arises, rather than retain a staff of writers (Yeoman 236).

In industry, most of the job openings for qualified technical writers and consultants lie in or near the major industrial cities, such as New York, Boston, Houston, the "Silicon Valley" area near San Francisco, and South Florida (VanAlstyne). This is mainly due to the fact that these areas play host to the majority of this nation's businesses and major industrial firms

Figure 12.17 Continued

Part VII of outline

9

who have the greatest need for the services of a professional technical writer.

According to authors Clarke and Root, "If our people will it, all our goals can be achieved with the help of science and technology. In that achievement, technical and science writers will be key participants" (Definitive Study 17).

Figure 12.17 Continued

Alphabetized list

Title is capitalized

WORKS CITED

Clarke, Emerson, and Vernon Root. A Definitive Study of Your Future in Technical and Science Writing. New York: Rinehart Rosen Press, Inc., 1972.

"Communications." Encyclopaedia Britannica. 1983 ed.

"Editorial Department, The." Publishers' Weekly 10 July 1981: 18.

Gould, J. R. and Wayne A. Losano. Opportunities in Technical Writing Today. Louisville: U.G.M. Publishing, 1976.

Pearsall, Thomas E. "Building a Technical Communication Program." ADE Bulletin Spring 1982: 15–17.

__________ and Donald H. Cunningham. How to Write for the World of Work. New York: Holt, Rinehart and Winston, 1982.

Schemehaur, P. J., ed. Writer's Market, 1983. Cincinnati, Ohio: Writer's Digest Books, 1983.

ThermoServ Company. A division of Dart and Kraft. Anoka, Minn. (a leaflet accompanying the product, n.d.).

United States Department of Commerce. The Statistical Abstract of the United States. Washington, D.C.: U.S. Government Printing Office, 1983.

Hanging indentation

Use of line where there is repetition of author

Titled page is numbered at bottom

10

Figure 12.17 Continued

11

VanAlstyne, Judith S. An interview on the subject of technical writing consultation in her office, Broward Community College, Fort Lauderdale, Fla., 12 March 1983.

Webster's Seventh New Collegiate Dictionary. 1980 ed.

Yeoman, William. "The Future of Technical Writing." In Jobs 1982–1983. New York: G. P. Putnam's Sons, 1982.

Figure 12.17 Continued

Former MLA Bibliography Style

The former MLA system requires a list of consulted references entitled "Bibliography," "Selected Bibliography," or "Works Cited." The titles "Bibliography" and "Selected Bibliography" indicate that the list includes background sources which are not actually referenced in the text. The former style of listing references differs from the new style in only a few ways. Entries for general books and reference works are the same. Magazine, journal, and newspaper entries differ as follows:

1. Magazine article (p. and pp. used for page references).

"Locking the Electronic File Cabinet." Business Week, 19 Oct. 1982, p. 124.

2. Journal with continuous pagination (comma rather than colon used after date).

Kolata, G. "Students Discover Computer Threat." Science, 2115 (1982), 1216.

3. Newspaper (p. and pp. plus column reference used).

O'Neill, Robert. "Computer-based Home Security Systems Sales Soar." New York Times, 15 Dec. 1983, p. 6, col. 1.

Former MLA Footnote Style

Under the former MLA system source references are inserted as footnotes or endnotes rather than as in-text, parenthetical notes. Raised Arabic superscripts are used in the text to indicate a source reference. The references are then typed either at the bottom of the appropriate page (footnotes) or on a separate page following the last page of the text. The notes are numbered sequentially throughout.

Under the MLA System a footnote or endnote for a book differs from a bibliographic entry in six ways:

1. Book with one author

Bibliography entry:

Bequai, August. Computer Crime. Lexington, Massachusetts: Lexington Books, 1978.

Footnote entry:

[1]August Bequai, Computer Crime (Lexington, Massachusetts: Lexington Books, 1978), p. 50.

First, the footnote entry is indented; second, the appropriate raised superscript number is included; third, the author's name is not transposed;

fourth, the punctuation tends toward commas rather than towards periods; fifth, the publication data is within parentheses; and sixth, a specific page reference is included.

The following is a guide to the former MLA System of footnoting:

2. Book with two authors

[2]Charles F. Hemphill and Robert D. Hemphill, Security Safeguards for the Computer (New York: Amocon, 1979), p. 72.

3. Book with three authors

[3]David K. Hsiao, Douglas S. Kerr, and Stuart E. Madnick, Computer Security (New York: Academic Press, 1979), pp. 201–202.

or

[3]David K. Hsiao and others, Computer Security (New York: Academic Press , 1979), pp. 201–202.

4. Edited book

[4]Roger C. Shank, Computer Models of Thought and Language, ed. Roger C. Shank and Kenneth Mark Dolby (San Francisco: W. H. Freeman, 1973), p. 528.

5. Essay in an edited collection

[5]Emmanuel LeRoy Ladurie, "The Historian and the Computer," in Territory of the Historian, ed. Emmanuel LeRoy Ladurie (Chicago: University of Chicago Press, 1979). p. 303.

6. Work in several editions or volumes

[6]Donn B. Parker, Crime By Computer, 2nd ed. (New York: Scribners, 1976), p. ix.

[7]Martin Anderson, Computer Networks, 2 vols. (Englewood Cliffs, N.J.: Prentice-Hall, Inc., 1981).

or

[8]Martin Anderson, Computer Networks (Englewood Cliffs, New Jersey: Prentice-Hall, Inc., 1981), II, 81.

(Notice that footnote 7 indicates the reference is to the works as a whole; therefore, no page number is included. Footnote 8 indicates that the reference is to material in the second volume, the Roman numeral, on page 81.)

7. Encyclopedia

[9]John Charles Smith, "Computers," Encyclopaedia Britannica, 1983.

8. Statistical abstract

[10]"Computer Services," Statistical Abstract of the United States, 82 (1979), 72.

9. Yearbook

[11]"Another Wire-transfer Theft," Facts on File, 1974, p. 239.

10. Magazines

[12]"Locking the Electronic File Cabinet," Business Week, 18 Oct. 1982, p. 124.

11. Journal with continuous pagination

[13]G. Kolata, "Students Discover Computer Threat," Science, 2115 (1982), 1216.

12. Newspaper article

[14]Robert O'Neill, "Computer-based Home Security Systems Sales Soar," New York Times, 15 Dec. 1983, p. 6, col. 1.

13. Bulletin

[15]Earl Brown, "Computer Laboratory Security Measures," Gainesville, Fla., 1984, p. 3.

14. Pamphlet

[16]National Center for Research in Vocational Education. Preparing for High Technology: Strategies for Change, Pamphlet No. 230 (Columbus, Ohio: Ohio State University, n.d.), p. 7.

15. Letter

[17]Information in letter to the author from Robert H. Larsen, President of R. H. Larsen and Associates of Fort Lauderdale, Fla., 11 Mar. 1984.

16. Interview

[18]Personal interview on microchip security with Arlen Lessin, President of Smart Card International of New York, 12 April 1984.

17. Table or illustration

[19]Charles L. Buckley, Accounting system flowchart, Introduction to Accounting (New York: Oxford University Press, 1980), p. 237.

Former MLA Subsequent Reference Style

Once the full reference data for a source (**a primary citation**) has been provided, a shortened form for all subsequent references to the same source (**a subsequent citation**) is employed. A subsequent citation usually consists of the author's last name and and reference. If no author is listed, the citation includes the article title or abbreviated article title and the page. If the author is named in the text, the note contains only the book or article title and the page reference. If more than one work by the same author is cited, the note contains the author's name, title of the work, and the page. If two entries by different authors with the same last name are used, the note includes the author's given name initial, the last name, and the page. Note the following sequence of primary and subsequent footnotes:

> [1]August Bequai, Computer Crime (Lexington; Massachusetts: Lexington, Books, 1978), p. 50.
>
> [2]Bequai, p. 201.
>
> [3]August Bequai, Computer Security (New York: Scribners, 1980), p. 28.
>
> [4]Bequai, Computer Security p. 30.
>
> [5]Marion Bequai, "Access Control Security," Time, 13 March 1981, p. 102.
>
> [6]M. Bequai, p. 308.
>
> [7]"Locking the Electronic File Cabinet," Business Week, 18 Oct. 1982, p. 124.
>
> [8]"Locking the Electronic . . . ," p. 127.

Some published works use the Latin abbreviations *Ibid.* (in the same place), *op. cit.* (in the work cited), and *loc. cit.* (in the place cited) for subsequent citations.

Figure 12.18 shows an endnote reference listing for the sample student research report as it would appear with a documented paper employing the former MLA style.

Title is capitalized.

ENDNOTES

Note identification.

[1]Webster's Seventh New Collegiate Dictionary. 1980 ed.

[2]Thomas E. Pearsall and Donald H. Cunningham, How To Write for the World of Work (New York: Holt, Rinehart and Winston, 1982), p. 1.

[3]"Communications," Encyclopaedia Britannica, 1983 ed.

[4]Judith S. VanAlstyne, an interview on the subject of technical writing consultation in her office, Broward Community College, Fort Lauderdale, Fla., 12 Mar. 1983.

Second reference to source

[5]Pearsall and Cunningham, p. 1.

[6]VanAlstyne.

[7]Emerson Clarke and Vernon Root, A Definitive Study of Your Future in Technical and Science Writing (New York: Rinehart Rosen Press, Inc., 1972), p. 32.

[8]Thermo Serv Company, a division of Dart and Kraft, Anoka, Minn. (a leaflet accompanying the product, n.d.).

n.d. means no date.

[9]"The Editorial Department," Publishers' Weekly, 10 July 1981, p. 18.

Titled page is numbered at bottom

12

Figure 12.18 Sample former MLA style endnote

14

[10]J. R. Gould and Wayne A. Losano, Opportunities in Technical Writing Today (Louisville: V.G.M. Publishing, 1976), p. 29.

[11]Thomas E. Pearsall, "Building a Technical Communications Program," ADE Bulletin, Spring 1982, p. 17.

[12]Pearsall, p. 17.

[13]VanAlstyne.

[14]P.J. Schemehaur, ed. Writer's Market, 1983 (Cincinnati, Ohio: Writer's Digest Books, 1983), p. 855.

[15]Schemehaur, p. 855.

[16]United States Department of Commerce, The Statistical Abstract of the United States (Washington, D.C.: U.S. Government Printing Office, 1983), p. 236.

[17]William Yeoman, "The Future of Technical Writing," in Jobs 1982–1983 (New York: G. P. Putnam's Sons, 1982), p. 236.

[18]VanAlstyne.

[19]Clarke and Root, p. 17.

Figure 12.18 Continued

EXERCISES

1. Locate the magazine and journal indexes in your library. Select an index appropriate to your field (*Applied Science and Technology Index, Business Periodicals Index, Criminal Justice Periodical Index, Hospital Literature Index, Microcomputer Index*, and so forth) and look up a general subject, such as nutrition, fire science, radiology, capital punishment, electronics, tourism, cardiology, dental hygiene, pollution, aviation, and so on. List five subclassifications which could be researched for development of a research report. Consider if the material is recent, not too technical, and available in sufficient amount.
2. Rewrite the following broad thesis statements to indicate a more limited focus and purpose.
   - **a.** Microchips have revolutionized the electronic industry.
   - **b.** Many new medicines are decreasing the risk of secondary coronaries in heart-diseased patients.
   - **c.** Macroengineering will improve continental travel in the future.
   - **d.** Fusion power will replace fission power.
   - **e.** The Twenty-first Century will usher in changes in the job market.
   - **f.** Fire fighting techniques are improving.
   - **g.** Family relationships are changing.
   - **h.** Computers are an aid to education.
   - **i.** Desalinization systems will provide more usable water.
   - **j.** Sun Belt cities are expanding rapidly.
3. Prepare six annotated bibliography cards on one of the above or other narrowed subjects. Include a general book, two reference books, two magazine or journal articles, and a newspaper article. Use the new MLA style unless otherwise directed by your professor.
4. Rewrite the following information in the MLA system "Works Cited" style.
   - **a.** An article in *Journal of Technical Writing and Communication* by Earl E. McDowell, L. David Schuelke, and Chew Wah Chung, entitled "Evaluation of a Bachelor's Program in Technical Communication," on pages 195–200 of volume 10 in 1980.
   - **b.** *Technical Writing: Structure, Standards, and Style,* by Robert W. Bly and Gary Blake, published in 1982 by McGraw-Hill Book Company in New York.
   - **c.** A *U. S. News and World Report* article, "When Job Training Will Be a Lifelong Process," on pages 25–26 of the May 9, 1983, issue.

d. An *Encyclopaedia Britannica* article, "Technical Communications," published in 1982, written by E. Charles Lloyd.

e. "Computer Industry Demands Writers" by David Hall, in the *New York Times*, page 4, column 3, on December 2, 1983.

5. Revise the following brief outline from a traditional, sentence format to a decimal, topic outline.

Thesis: Striking changes in transportation modes are predicted for the next century.

- **I.** Cars will still be the primary means of getting around.
  - **A.** Cars will be totally electronic.
    - **1.** Voice commands will turn the car on and off, select radio channels, and activate the interior ventilation system.
    - **2.** All vehicles will be equipped with a radio-telephone.
    - **3.** Electronics will control fuel use.
  - **B.** Design will be streamlined.
    - **1.** Future cars will be low-slung, with sharply raked windshields.
    - **2.** Undercarriages and wheel wells will be enclosed.
    - **3.** Windows will be flush with the body.
    - **4.** Grilles and outside ornament will disappear.
  - **C.** Fuel consumption will improve.
    - **1.** Fuel mileage will improve to 75 to 100 mpg by 2010.
    - **2.** Engines will be smaller and constructed of molded ceramics to descrease weight.
    - **3.** Electronic fuel injection and transmission will maintain engines at near constant speed.
    - **4.** More cars will be fueled by methanol fuel made from coal.
    - **5.** Eventually cars will run on hydrogen and be able to travel on both water and land.
- **II.** Rapid-transit systems will ease congestion in cities and metropolitan areas.
  - **A.** Urban rail systems will expand between cities.
  - **B.** Double-deck, multisectional buses will carry as many as 150 people on separate busways.
    - **1.** Sleeper seats, videograms, and meals will be standard.
    - **2.** Fare collection will be simplified by electronically coded cards and single monthly bills.

C. Moving sidewalk belts will glide along elevated guideways at speeds of 10 to 15 mph.

D. Rail travel will improve.

1. "Bullet" passenger trains will travel between major urban areas at 160 mph.
2. Magnetic-levitation trains which can travel at more than 250 mph are being tested in Japan.

III. Future generations will witness new ventures in the sky.

A. By the year 2000 jetliners will carry as many as 1000 people.

B. One-passenger flying machines are being tested at Lockheed.

6. Read the following excerpt from an article on innovative computer software programs:

Windows on the World

Two software leaders give video screens a new look

Like women's fashions, computer buzz words change with the season and tend to hide more than they reveal. Last year's programs were all "user friendly," although many proved painfully difficult to master. This year, software is "integrated," which means that information from one program can sometimes be merged with data from another. Industry watchers are now getting a preview of the pet phrase for 1984. Two leading computer software companies, Microsoft and VisiCorp, are offering products with "windows," a system that lets users run several different programs at once, each displayed in a separate section of the video screen.

VisiCorp, the San Jose, Calif.-based publishers of the successful VisiCalc program for financial analysis, next month will begin shipping its VisiOn windowing package. Meanwhile, Microsoft, the leading personal computer software publisher (1983 sales: $100 million), has unveiled a competing product called Windows. Said Microsoft Chairman Bill Gates: "This is a milestone in software."

Each program uses a cigarette pack-size "mouse" as a control device; each allows users to split their screens into rectangular blocks, or windows, giving them the look of desktops littered with sheets of paper. Both systems attempt to address two fundamental challenges facing the personal computer industry: how to get the same program to run on machines put out by different manufacturers, and how to swap information smoothly between different programs. At present, for example, software for an IBM machine will not run on an Apple computer, and most users cannot easily take information from a financial analysis program and send it to a client via electronic mail.

a. Write a summary note of the entire excerpt.

b. Write a paraphrase note of the second paragraph.

c. Write a quotation note (full or partial sentence) of the first sentence of the third paragraph.

WRITING OPTIONS

1. Research and write a documented report on careers in your field. Include a background review, optional opportunities, education requirements, salaries, details of the work performance activities, locations, numbers in the field, prospects for the future.
2. Research and write a documented report on a breakthrough discovery in your field. Investigate who made it, how it was made, what are its uses, what benefits does it offer, and what problems does it pose.
3. Research and write a documented report on some aspect of the Twentieth Century. Following is a list of broad subjects which can be narrowed to focused topics:

the American family	role of the Far East
education	transportation
health care	fuels
medicine	travel modes
crime	satellites
life expectancy	electronic marvels
food production	genetic engineering
minerals and materials	weather modification
pollution	architecture
the shape of cities	home appliances
the role of computers	population
the economy	space exploration
new occupations	other?

13

ORAL COMMUNICATIONS

Skills:

After studying this chapter, you should be able to

1. Identify potential professional situations in which oral communication skills are necessary.
2. Prepare for and role play job interviews with appropriate actions, responses, and questions.
3. Conduct professional transactions on the telephone.
4. Prepare for informal oral communication situations, such as group discussions, information gathering, appraisals, and reprimands.
5. Analyze nonverbal messages and respond to them appropriately.
6. Analyze, organize, prepare, and deliver a persuasive or informative oral report with appropriate visual materials.

INTRODUCTION

Effective oral communications are important in the work setting. Every organization requires its employees to develop speaking skills. To gain employment you must be able to speak persuasively about your own potential as well as express yourself knowledgeably about the organization which has granted you an interview. Secondly, once you are employed, you will be expected to demonstrate careful interpersonal communication skills in a variety of tasks ranging from telephone transactions and group decision-making discussions to formal speech presentations. You should even be aware of the nonverbal communications you will send and receive.

INTERVIEWS

Chapter 3 covers the writing strategies of the resume and cover letter necessary to obtain an interview. Speaking skills are equally important once the interview has been granted.

Preparation

There are many things you can do to prepare yourself for an interview.

1. Determine if the interview is a **screening interview** or a **line interview.** In a screening interview, numerous applicants are invited to speak with a number of interviewers, perhaps a search committee as well as the officers and directors of an organization. The objective is to narrow the pool of candidates for more intensive, follow-up interviews. In the line interview, only a few select applicants will be intensively interviewed by fewer personnel. The screening interview will be more informal than the line interview, yet in both situations what you say and how you say it will be assessed carefully.
2. Learn about the organization and the job. Speak to present employees, professors, and professionals who have contact with the organization. Know what services or products the company offers. Review its annual report if possible. Find out as much as you can about employment policies, potential layoffs or transfers, opportunities for advancement, union relations, competitive situations, and on-the-job training opportunities. Ask for a job description in advance to determine exactly what skills are required for the position.
3. Review your resume. Prepare brief, oral answers to the questions which are likely to be asked and prepare your own questions. This chapter will provide lists of potential questions.

At the Interview

Getting started. Although the interviewer has set the time, place, and type of interview and will control the line of questioning, you, the interviewee, have power to control certain factors which will put you at ease in order to present yourself at your best. You have alternatives, such as suggesting alternate times for the interview, asking for clarification of questions, taking your time in responding, refusing to answer personal questions which are not relevant, and asking questions of your own.

Body movement. Upon entering, pay attention to your posture and bearing. Shake hands firmly. If you are a woman, offer your hand first, for many males are still uncertain if it is proper etiquette to initiate a handshake with a woman. A handshake establishes confidence and warmth.

Sit comfortably without slouching or appearing too stiff. Keep both feet on the floor, and lean slightly forward occasionally to project poise and interest. Control your nervousness by avoiding any fidgeting or fiddling with pens, hair, papers, purses, briefcases, and the like. Do not drum your fingers or tap your feet.

Personal space. Establish a distance between you and your interviewer which is close enough to convey warmth and sincerity yet far enough to convey a degree of formality. Avoid sitting side-by-side. Do not violate the interviewer's space by sitting too close, moving behind the desk, touching beyond the initial handshake, or touching objects on the desk.

Eye contact. Look directly at your interviewer. Eye contact suggests confidence, honesty, and interest. Watching the other person will reveal how that person is reacting to your answers. You can expand, modify, or cut short your responses. Control your own facial expressions. Look alert and be responsive to the discussion. Try to judge the impression you are making during all stages of the interview.

Voice. The acoustics of the room, the distance from the interviewer, and your own anxiety should be noted in order to modulate your voice to create a friendly, yet assertive, impression. If you talk too loudly or too high pitched, you will be offensive and boorish. If you talk too softly, you will appear "wishy-washy" or uninterested.

Accept or ask for water or coffee if your voice becomes raspy. A few sips will relax your vocal cords.

Questioning Skills

Your resume and application will have presented basic information about you, but both your interviewer and you will ask questions. The interviewer will expect you to summarize your qualifications. In addition, the employer wants to learn about your potential, your expectations, your attitude toward the organization, and your personality and character. By thinking about the following questions which you may be asked, you can prepare oral responses which are complete and confident.

Interviewer's Questions

1. Can you tell me about yourself? Expand on your resume? (Give a summary of your employment, educational, and personal background.)
2. Why do you believe you are qualified for this position? (Stress skills.)
3. Why did you select us in your job search? (Demonstrate that you have done your homework about the organization.)
4. What are your long-term career goals? What do you want to achieve in five years? Ten years? (Demonstrate that you have goals and expectations which are realistic.)
5. What are your strongest (weakest) personal qualities? (Name two or three each; negatives can be balanced against positives or recognized with a goal for improvement.)
6. How do you handle stress? How would you handle conflict between yourself and a subordinate (supervisor, peer)? (Summarize how you cope with personal stress, such as physical exercise,

reading a book, gardening, and the like. Be prepared to answer how you approach problems.)

7. What have been your most satisfying and most disappointing school or work experiences? (Recite your accomplishments and stress how you overcame disappointments.)
8. What are your attitudes toward unions (absenteeism, punctuality, geographical transfers, shift work, weekend assignments, travel)? (Know in advance what the organization expects.)
9. Do you have any home or personal problems which may bear on your performance? (Be honest while stressing how such situations are being modified.)
10. What salary do you have in mind? Would you accept x amount? (Be prepared to give a minimum or a range, emphasizing need and goals. Accept or ask for clarification about work hours, promotions, and raise potential.)

Throughout the interview you should feel free to ask questions yourself. Seek clarification of any vague questions asked of you. Let your own line of questioning reflect that you are interested and serious about the organization. Decide which of these typical questions you want answered.

Interviewee's Questions

1. What are the organization's long- and short-term plans? Is the organization expanding or shrinking?
2. What is the potential for advancement in this position?
3. How does your organization support employee advancement? Are there educational opportunities? Grants? Training programs? Performance reviews?
4. What is the leadership structure? Will I have any decision-making authority?
5. What are the organization's relations with the community? Consumers? Related organizations?
6. What are the grievance procedures? Lay-off policies? Leave policies?
7. What are your hours of operation? How much travel is expected? Do employees normally work many hours of overtime?
8. Are there any benefit provisions? (Insurance, pension plans, stock options, cars or housing provided, travel per diem, employee discounts, and the like.)
9. How is the housing market in this area? What cultural, recreational, and social opportunities are available in this city? How would you rate public transportation, schools, municipal services?
10. What are the wages and salaries? Are increases based on performance or indexes? What are the means of advancement?

Ending the Interview

The interviewer will usually signal the end of the interview, but you should have some closing questions. Ask what further information you should supply, what follow-up steps will be appropriate, and how and when you will learn the outcome. Remember that your interviewer is a busy person with others to interview and additional tasks to perform. Do not dawdle, but thank the person for the time and interest provided to you and leave decisively.

TELEPHONE CONVERSATIONS

Once you have the job, you become a representative of the organization. Many transactions are conducted by telephone both within and without the company. Your telephone etiquette reflects not only your own oral skills but also the organization's image.

Certain basic strategies will help you to convey and to receive messages in a professional manner. Talking on the phone involves both talking and listening skills. Pay attention to the volume, tone, and clarity of your voice. Do not shout or whisper. Your mouth should be about an inch or two from the speaker. Your tone should be warm and pleasant. Enunciate clearly; avoid slurring of "Hullo," "Whajasay?" "Yeah," and the like. Avoid distracting background noises, such as another conversation, typewriters, radios, tapping, blowing, or chewing.

Be prepared to take notes. Have a notepad, pens, and pencils on hand near your telephone. Have on hand, too, any reports, letters, or other printed matter which may be references for your conversation. Nobody wants to wait while you put down the telephone to search for these things.

Placing a call. When placing a call, identify yourself by a courteous "Hello" and state your name and other identifying comments, such as "Hello, this is Judy Withrow, the training consultant with Writing Skills Management. I spoke to you last Friday about the possibility of conducting a seminar for your midlevel management people." After an acknowledgment, proceed to state the purpose of your call, such as "I'm calling to determine if you received the brochures I mailed to you and to set a date for an interview." After transacting your business, specify what it is you want your listener to do: call you back in an hour or a week, mail you material, or confirm details in writing. If it is appropriate, thank the listener for the information, interest, or time.

Receiving a call. When receiving a call, say "Hello" courteously and identify yourself, your department, or your organization: "Hello, Tom Brown speaking," "Hello. Personnel Department, Miss Shipley speaking," or "Ace Company. This is Bob Martin, Sales Manager. May I help you?" End the conversation courteously, too. It may be appropriate to say, "Thank you for calling," "I'll send the materials to you in today's mail," or "I hope I've assisted you." Say "Goodbye" in response to the caller's closing, and hang up carefully. A receiver which is banged down or dropped on its cradle may unduly irritate the caller who has not yet hung up.

If you answer the telephone for someone else, be doubly courteous. Let the caller know that the right number has been reached: "Hello, Ms. Steven's office; this is Jerry Thomas, her assistant speaking. May I help you?" Do not make lame excuses for another's absence, such as "She's busy right now" or "I don't know where she is." State that the individual is engaged, in conference, or out to lunch, and offer to take a message. Do whatever you can to have the call returned.

OTHER ORAL COMMUNICATIONS

As a professional worker you will be expected to take part in a variety of oral information situations. These may include group or panel discussions, information gathering interviews, and appraisals and reprimands.

Group Discussions

From time to time you will be part of a decision-making group—a committee, a department, or even a larger group. Large groups tend to operate through parliamentary procedure to reach decisions while small groups operate less formally. We are concerned here only with small task groups.

A number of factors affect group processes, such as the purpose for which the group exists, the personal goals of its members which may be in conflict with the overall purpose, the permanency of the group, the power and relationships of the members of the groups, and the methods used for deciding. The most effective group is one with a clear purpose which is understood and supported by all of its members. Further, the longer a group can work together the less it tends to be dominated by one or two people. And, finally, a group which knows its decision will be accepted by others will be more successful than a group that perceives an outside threat or overriding decision-maker.

Decisions are made in groups by *majority rule, compromise,* or *consensus*. Majority rule, a decision by a vote, is common but tends to polarize the two camps making the losers less committed or even antagonistic to the decision. In a compromise decision, both sides give up a little in order to gain a little. Collective bargaining uses compromise decision-making methods. Compromise is effective if members of the group have to answer to larger constituencies. The constituencies can feel they won something and were not total losers, yet since nobody is 100 percent pleased with the decision, members may not work very hard to implement the decision. In consensus decisions all members agree on the decision. It is the most difficult way to decide but the most effective because all members are satisfied, which will lead to better productivity and commitment to implementation of the decision.

In order for a group to reach consensus decisions, all members should participate in clarifying a goal and establishing procedure. Establishing an agenda will facilitate consensus. If you are the initial leader of the group, you may exert a positive influence. Let each member talk freely about the goal and suggest procedures for the group to follow. Next, encourage or participate in a group discussion to determine the status and causes of the problem or situation. If appropriate, an effective group will then set criteria for evaluating solutions or decisions. Brainstorming solutions is a creative way to consider all possible solutions. In the brainstorming strategy all members offer as many different solutions as possible with no interrupting criticism or evaluation. These suggestions should be listed on a chalkboard or flip chart. Next, the solutions are weighed against the criteria to determine which is best in terms of effectiveness, cost, simplicity, possible implementation, and acceptability to the group. Finally, the group should decide how to implement the solution, delegating and accepting responsibility.

Information Gathering

You may often find yourself acting as an interviewer to obtain information from your co-workers, superiors, and people in other organizations. Do not waste time. Determine in advance exactly what you need to know, and organize and jot down topics which need to be covered. Be prepared to take notes.

In the actual interview clarify your overall purpose for the inquiry. Ask clear questions. Keep each question brief so as to elicit specific responses. Give the person plenty of time to answer. Be courteous and express your appreciation for the assistance. Finally, offer to share the results of your information gathering.

Appraisals and Reprimands

You will probably meet with your supervisor from time to time to assess your job performance. As you move up the career ladder, it may become your responsibility to appraise the performance of your subordinates. Giving and receiving reprimands are also inevitable transactions in any organization. Both appraisals and reprimands tend toward emotional communication, so strategies should be aimed toward objectivity.

A performance appraisal interview should be scheduled in advance so that both sides can prepare. The interview should be limited to a specific time period—15 to 30 minutes—to avoid repetitious harangues. Ideally, both parties should prepare separate written assessments of the performance. With these exchanged in advance, the parties may enter the actual interview prepared to concentrate on task-related factors rather than personalities. Explanations for poor performance may be offered, but the discussion should center on goals and methods for improvement.

In the reprimand situation, objectivity is also the key. One should offer written policies or procedural manuals which spell out the expected activity to substantiate that a violation occurred. The supervisor should attempt to determine if the violation was due to lack of information or was willful. Clear communication is essential. The supervisor needs to ask questions, use feedback, and be sensitive to the feelings and possible penalty to the violator. The violator must be honest and concerned. Questions about recourse, appeals, and corrective actions are appropriate.

NONVERBAL MESSAGES

Oral communications involve more than the words being spoken. Our senses of sight, hearing, touch, and smell are also receiving messages which can color our perceptions of the spoken word. These perceptions are called *nonverbal messages*. A number of the message carriers have already been mentioned—posture and position, voice modulation, eye contact, handshaking, background noise, and the like. For effective communications you should be alert both to the nonverbal messages you send and those you receive. Ideally, nonverbal messages will be consistent with verbal messages, but such consistency is often violated.

Overt actions. Actions often do speak louder than words. Obvious actions, such as jabbing with a finger, pacing back and forth, waving papers around, and pounding on desks quite obviously signal aggression and anger. But subtle body actions, such as slouching, staring at the ceiling, standing with crossed arms, and the like, convey messages, too. Slouching conveys boredom; staring at a ceiling conveys disinterest; and crossed arms convey coldness or aggression. Open arm and body positions convey trust, warmth, and sincerity.

Covert actions. Some actions, such as blushes, tight jaws, smiles, and the like, reveal how a person is feeling. Who has not been confused by the person with furrowed brow and a tight-mouthed smile asserting,

"No, I'm not mad at you"? Be alert to these signals in interpersonal communications, and learn to control your own covert actions in order to convey consistent verbal and nonverbal messages.

Eye contact. Eyes are the primary conveyers of nonverbal messages. Direct eye contact conveys interest and receptivity while reduced eye contact may convey concentration or deceit. Between communicators eye contact regulates interactions and monitors the verbal messages. Use your eyes to your advantage.

Appearance. Physical appearances also convey messages although often inaccurately. We tend to stereotype persons according to three basic body types: athletic, frail, or obese. We expect the athletic person to be strong, mature, or self-reliant while we expect the frail person to be indecisive, bookish, or pessimistic. Examine your personal tendency to stereotype people by their body type and actively rethink the validity of these prejudices. Be aware of how your own body type influences the perceptions of others. Changeable factors in appearance—length of hair, mustaches, eyeglasses, uniforms, jewelry, cosmetics, hem lengths, and so on—convey impressions too. We attribute skills, personality traits, and abilities based on how people look.

Space. Space is also an important nonverbal message carrier. The arrangement of office furniture or conference facilities can enhance or detract from open communications. Also individuals have a sense of territory; to encroach on another's space is perceived as a threat or an aggression. Even the distance that we set between ourselves and those with whom we are communicating conveys messages. Arabs and Latins tend to communicate at much closer distances than do North Americans or the British. Sensitivity to space arrangements will aid you in communicating the messages you intend.

ORAL REPORTS

You will be called upon in your career to make any number of informal or formal oral reports. These may be reports at meetings of your peers and supervisors, presentations at training seminars, speeches at conventions or before civic groups, or presentations to explain proposals and other projects. Oral reports are classified as:

- Impromptu speeches
- Memorized speeches
- Manuscript speeches
- Extemporaneous speeches

The impromptu speech is an off-the-cuff presentation likely to occur at a meeting where you are either asked about a project or you decide to present your views on an agenda item. A memorized speech may be appropriate for material which must be communicated many times, but it is difficult to avoid sounding wooden and mechanical in memorized speeches. A manuscript speech, one which is read, may be appropriate to convey very technical or detailed information but calls for exacting practice to avoid a monotone delivery and lack of eye contact. The extemporaneous report is the most widely used and effective oral presentation.

EXTEMPORANEOUS REPORTS

The extemporaneous report requires careful audience analysis, clear purpose, logical organization, supportive visual materials, sufficient rehearsal, and skillful delivery.

Audience

Previous chapters have stressed that an analysis of your audience is essential for effective written reports. The public speaker must also consider the audience who will listen to the oral report. Besides thinking about how much the audience already knows about your subject, what level of technical language is appropriate, and your relationship to the group, you should seek to discover in advance the average age of the group, political persuasions, religious or ethnic affiliations, rural or urban interest, and sex. Knowledge of these factors will help you to infer how the listeners will receive your information. These factors should indicate the degree of formality appropriate to your talk.

It is equally important to continue analyzing your audience during your speech through feedback. You can gauge your audience's reactions to your speech by noting facial expressions, postures, applause, and the like. The alert speaker will make adjustments in delivery based on this feedback.

Purpose

It is important to have a clear purpose in mind: to entertain, to persuade, or to inform. We are not concerned here with the entertaining speech, but as a professional many of your reports will persuade or inform your audiences.

Persuasion. The persuasive speech attempts to bring about overt action or to change beliefs and attitudes. Examples of overt actions are the purchase of products, the election of officers, the adoption of policies, or the alterations of procedures. A persuasive speech may also change workers' attitudes toward minority and women workers, encourage pride in organizational membership, and so forth.

Three strategies will help you to deliver an effective persuasive speech. First, consider *credibility*. Arrange to give your speech in a setting which is comfortable for the audience. Arrange, also, to be introduced by a person who is well-liked by the group and who will stress your qualifications to speak on the subject. Second, establish *identification* with the group by the way you dress and act or by actually expressing similarities in ideas, beliefs, or experiences. State your goals in an honest, friendly, and assertive manner emphasizing the rewards (anything that meets the needs and desire of the group) that your listeners will receive if they are persuaded. Finally, be prepared to present not only your evidence for your conclusions but also the *warrants* (reasons) for believing the evidence is relevant to the conclusions.

Information. The informative speech is the most common among professionals. Such reports may be patterned along the same lines as written reports, such as progress reports, instructions, analyses of processes, descriptions of mechanisms, and so forth. Do not lose sight that your primary goal is to impart information.

Organization of Report

Your evidence or data must be logically organized. A listener is not a reader. A listener cannot back up to review information or skip ahead to the conclusion. Organize your report with the listener in mind.

If your purpose is to persuade, you may consider the problem-solution organizational approach or the advantages-disadvantages approach. Figure 13.1 illustrates the overall organization of these approaches.

If your purpose is to inform, you will organize along the lines of the type of report which you are presenting orally. Figure 13.2 illustrates an overall organization approach to your speech.

Outlining. Next prepare an outline of your actual speech and gather your information. Other chapters in this book review outlines and organization for specific types of reports.

Notecards. Third, using your outline, prepare 3 by 5-inch or larger notecards of the main topics of your speech. Underline key points in red, and indicate by asterisks where you plan to use your visual materials. Use only one side of the notecards, and do not overload a card.

PROBLEM-SOLUTION APPROACH

I. Approach
   A. Gain attention and goodwill.
   B. Develop credibility, if necessary.
   C. Orient receiver to subject and purpose.

II. Body
   A. Develop problem.
      1. Explain symptoms or results (problem description).
      2. Explain size and/or significance.
      3. Explain cause.
   B. Develop solution.
      1. Explain solution.
      2. Explain how solution eliminates problem.

III. Conclusion
   A. Appeal for action or desire belief.
   B. Allow for discussion.

ADVANTAGES-DISADVANTAGES APPROACH

I. Introduction
   A. Gain attention and goodwill.
   B. Develop credibility, if necessary.
   C. Orient receiver to subject and purpose.

II. Body
   A. Explain disadvantages of present situation.
   B. Explain advantages of new idea, proposal, policy, or situation.
   C. Explain that advantages cannot be obtained without proposed changes.

III. Conclusion
   A. Appeal for action.
   B. Allow for discussion.

Figure 13.1 Persuasive oral report approaches

INFORMATIVE SPEECH APPROACH

I. Introduction
   A. Gain attention and goodwill.
   B. Orient receiver to subject and purpose.

II. Body
   A. Present data in logical organization.
   B. Repeat key terms and provide verbal transitions between parts.
   C. Employ visual aids.

III. Closing
   A. Summarize.
   B. Emphasize.
   C. Allow for questions and/or discussion.

Figure 13.2 Informative oral report approach

Visual Material

Prepare visuals to clarify and emphasize your information. Visual materials may include chalkboard or flip-chart notes and drawings; posters of tables, charts, drawings, and the like; handout sheets; exhibits of models, equipment, or brochures; dramatizations and demonstrations; and/or projections of slides, filmstrips, or transparencies.

The size of the room or auditorium, the kind of people in your audience, the available monies, the available equipment (chalkboards, projectors, tables, and so forth), and the nature of your speech are all factors for consideration in planning visual material.

Chapter 4 discusses the value and conventions of graphics in written reports. Those same principles govern visual materials for oral reports.

Posters and transparencies. Two simple graphic materials to construct are posters and transparencies. Posters do not require special equipment. They may be mounted on a wall, blackboard, or tripod near your podium for reference during your speech. Use a pointer to direct attention to visual detail. Be careful not to block the content with your body. Allow your audience to examine the supplements following the report.

Transparencies will require an overhead projector and a screen or blank wall. You may cover some information with paper and then expose the data as you are ready to discuss a point. You may add notations or color to transparencies during your speech without losing eye contact. In addition, your artwork can be photocopied for handouts to relieve your audience from copying complex graphics.

Effective design. Effective visual materials are characterized by five factors:

- Simplicity
- Unity
- Emphasis
- Balance
- Legibility

1. **Simplicity.** Fewer elements are more pleasing to the eye than are a hodgepodge of detail. Bold, key detail has more impact than does complex art. If the material contains verbal data, limit it to 15 to 20 words.
2. **Unity.** Unity may be achieved by overlapping elements, arrows, and a conformity of shapes and sizes on any one graphic. Do not lay out your drawings randomly.
3. **Emphasis.** Use color, bold sizes, and white space to achieve emphasis.
4. **Balance.** Balance may be formal or informal. Formal balance is achieved when an imaginary axis divides the design into two mirror halves horizon-

tally or vertically. Informal balance is asymmetrical. It is more dynamic and more attention-getting.

5. **Legibility.** Wording on visual materials should be minimal. Sanserif or gothic letters are easier to read than script. Use capital letters for short titles and labels, but use a combination of upper- and lowercase letters for verbal content of six or more words. Base your letter size on the maximum anticipated viewing distance. The maximum viewing distance is generally accepted as being eight times the horizontal dimensions of the graphic. Thus, a 2-foot-wide poster has a maxium viewing distance of 16 feet, and a transparency projection on a 4-foot screen has a maximum viewing distance of 32 feet. Minimum letter sizes are 1-inch high on posters and ¼-inch high for transparencies. Thick letters are more legible than thin letters.

Rehearsal

Rehearse your speech a number of times before you actually give it. Practice before a mirror, and use a tape recorder. If possible, give your speech to a small group of friends. Ask them to assess your poise, eye contact, voice, gestures, and rate.

Your voice should be conversational, confident, and enthusiastic. Avoid a monotonous sound by varying the pitch, intensity, volume, rate, and quality of your voice.

Delivery

All of the preceding steps should prepare you for an effective delivery. How your audience perceives you and your information depends on skillful delivery.

General appearance. Dress appropriately. Approach the podium confidently and maintain good posture. Gesture naturally. Gestures may be larger in a large room than in a small room. Be animated and maintain eye contact. Do not be ramrod stiff nor clutch the podium. Do not jingle keys and the like or make distracting adjustments to your hair, glasses, or clothes.

Audience interaction. Pause before you begin your speech to gain the listeners' attention. Begin forcefully and engagingly. Keep tabs on your audience. Your listeners will be confirming or contradicting what you say through nonverbal messages. Allow for questions and discussion at the end of your report.

Notes and visual usage. Use your notes and visuals with ease. Place these items in comfortable positions. Do not fidget with your cards or pointers. Stand aside when referring to visual materials and maintain eye contact as you make your points about the materials.

Voice. Try for vocal variation. Let your voice exude warmth and sincerity. Pronounce your words clearly and distinctly. Do not vocalize "uh's," "um's," and other nervous sounds. Pause when appropriate.

Oral Rating Sheets

Figures 13.3 through 13.6 show speech rating blanks. Figure 13.3, a general speech rating blank, may be used to assess any extemporaneous speech. Figures 13.4, 13.5, and 13.6 may be used to assess an oral description of a mechanism, instructions, or an analysis of a process, respectively.

EXTEMPORANEOUS SPEECH
RATING SHEET

Speaker ______________ Subject ______________ Evaluator ______________

ITEMS	COMMENTS	SCORE
ORGANIZATION: Clear arrangement of ideas? Introduction, body, conclusion? Pattern of development adapted to ideas and audience?		
LANGUAGE: Clear, accurate, varied, vivid? Appropriate standard of usage? In conversational mode?		
MATERIAL: Specific, valid, relevant, sufficient, interesting? Properly distributed? Adapted to audience? Personal credibility? Use of evidence?		
DELIVERY: Poised, at ease, communicative, direct? Eye contact? Aware of audience reaction to speech? Do gestures match voice and language?		
ANALYSIS: Approach to subject original, interesting? Central idea, purpose clear, divided into significant, interesting, subordinate ideas?		
VOICE: Pleasing, adequate, distracting? Varied or monotonous in pitch, intensity, volume, rate, quality? Expressive of logical emotional meanings?		

TOTAL ______

SCALE:

10	7	4	1
Superior	Average	Inadequate	Poor

Figure 13.3 Extemporaneous speech rating sheet

MECHANISM DESCRIPTION RATING SHEET

Speaker __________ Instructions __________ Evaluator __________

Did the speaker	Yes	Somewhat	No	Comment
1. Make necessary preparations before starting?				
2. Define intended audience?				
3. Name mechanism precisely?				
4. Define and/or state purpose of mechanism?				
5. Provide an overall description?				
6. Discuss the operational theory?				
7. State by whom, when, and where the mechanism is operated?				
8. Provide a list of the main parts?				
9. Describe the parts in the order listed?				
10. Define and/or state purpose of each part?				
11. List subparts of assemblies?				
12. Describe each part adequately?				
13. Avoid wordiness?				
14. Assess the advantages and disadvantages?				
15. Describe optional uses?				
16. Compare the mechanism to other models?				
17. Discuss the cost and availability of mechanism?				
18. Avoid reading the presentation?				
19. Use adequate, well-organized notecards?				
20. Show evidence of rehearsal?				
21. Maintain effective eye contact?				
22. Speak clearly so all could hear?				
23. Use appropriate graphics?				
a. Number and title clearly?				
b. Print legibly?				

Figure 13.4 Rating sheet for oral description of a mechanism

Did the speaker	Yes	Somewhat	No	Comment
c. Keep graphics simple and uncluttered?	___	___	___	___
d. Lay out logically?	___	___	___	___
e. Credit sources of graphics?	___	___	___	___
24. Handle graphics with ease?	___	___	___	___
25. Do you feel you can judge the reliability, practicality, and efficiency of the mechanism?	___	___	___	___

SUGGESTED GRADE ___________

Figure 13.4 Continued

INSTRUCTIONS RATING SHEET

Speaker ____________ Instructions for ____________ Evaluator ____________

Did the speaker	Yes	Somewhat	No	Comment
1. Make necessary preparations before starting?	___	___	___	___
2. Define intended audience?	___	___	___	___
3. Provide a specific, limiting title?	___	___	___	___
4. State the instructional or behavioral objective?	___	___	___	___
5. Stress the importance of the instructions?	___	___	___	___
6. Define key terms?	___	___	___	___
7. State preliminary warnings or cautions?	___	___	___	___
8. Divide the process into main steps?	___	___	___	___
9. Provide a precise list of tools, materials, etc.	___	___	___	___
10. Divide the main steps into substeps?	___	___	___	___
11. Provide warnings and notes for individual steps?	___	___	___	___
12. State each step as a command?	___	___	___	___
13. Avoid combining steps and substeps?	___	___	___	___
14. Express all steps in parallel, grammatical terms?	___	___	___	___
15. Avoid wordiness?	___	___	___	___
16. Avoid reading presentation?	___	___	___	___
17. Use adequate, well-organized notecards?	___	___	___	___
18. Show evidence of having rehearsed?	___	___	___	___
19. Maintain effective eye contact?	___	___	___	___
20. Use appropriate graphics?	___	___	___	___
a. Number and title clearly?	___	___	___	___
b. Print legibly?	___	___	___	___
c. Keep graphics simple and uncluttered?	___	___	___	___
d. Lay out logically?	___	___	___	___

Figure 13.5 Rating sheet for oral instructions

Did the speaker	Yes	Somewhat	No	Comment
e. Credit sources of graphics?	____	____	____	____________
21. Handle graphics with ease?	____	____	____	____________
22. Invite questions from the audience?	____	____	____	____________
23. Do you feel you could perform the set of instructions accurately and efficiently?	____	____	____	____________

SUGGESTED GRADE ____________

Figure 13.5 Continued

PROCESS ANALYSIS RATING SHEET

Speaker ____________ Process ____________ Evaluator ____________

Did the speaker	Yes	Somewhat	No	Comment
1. Make necessary preparations before starting?	___	___	___	________
2. Define intended audience?	___	___	___	________
3. Name process precisely?	___	___	___	________
4. Select a process primarily involving human action?	___	___	___	________
5. Define or state purpose of process?	___	___	___	________
6. Define or explain terms?	___	___	___	________
7. Explain theory on which process is based?	___	___	___	________
8. Indicate by whom, when, and where the process is performed?	___	___	___	________
9. Indicate special conditions, requirements, preparations, precautions which apply to the entire process?	___	___	___	________
10. Precisely list materials, tools, and apparatus?	___	___	___	________
11. Divide process into five or six main stages?	___	___	___	________
12. Arrange steps in numbered, chronological order?	___	___	___	________
13. Provide a flow chart of main steps?	___	___	___	________
14. Discuss steps in order listed?	___	___	___	________
15. Define or state purpose of each main step?	___	___	___	________
16. Divide main steps into substeps?	___	___	___	________
17. Describe special conditions for each step?	___	___	___	________
18. Explain theory which applies only to one step?	___	___	___	________
19. Adequately analyze each main step with emphasis on why, to what degree, to what extent, etc.?	___	___	___	________

Figure 13.6 Rating sheet for oral analysis of a process

Did the speaker	Yes	Somewhat	No	Comment
20. Avoid excessive detail?	____	____	____	____
21. Evaluate effectiveness of process?	____	____	____	____
22. Discuss advantages/disadvantages?	____	____	____	____
23. Discuss importance of process?	____	____	____	____
24. Evaluate results of process?	____	____	____	____
25. Explain how process is part of larger process?	____	____	____	____
26. Compare process to similar processes?	____	____	____	____
27. Assess cost and time factors?	____	____	____	____
28. Avoid reading presentation?	____	____	____	____
29. Use adequate, well-organized notecards?	____	____	____	____
30. Show evidence of rehearsal?	____	____	____	____
31. Speak clearly so all could hear?	____	____	____	____
32. Avoid shifting to instructions?	____	____	____	____
33. Use appropriate graphics?	____	____	____	____
a. Number and title clearly?	____	____	____	____
b. Print legibly?	____	____	____	____
c. Use color effectively?	____	____	____	____
d. Keep graphics simple and uncluttered?	____	____	____	____
e. Lay out logically?	____	____	____	____
34. Handle graphics with ease?	____	____	____	____
35. Invite questions from the audience?	____	____	____	____
36. Do you feel you can judge the reliability, practicality, and efficiency of the process?	____	____	____	____

SUGGESTED GRADE ____________

Figure 13.6 Continued

EXERCISES

1. Working in pairs, devise interview questions to use both as interviewer and interviewee for a specific type of position at an appropriate fictitious company. Role play in front of the class the interview situation. The class will assess your verbal and nonverbal communication skills.
2. Working in pairs, role play a business telephone transaction. Devise your own transaction or try the following situation. The caller has mailed a cover letter and resume to the director of personnel. The caller is trying to find out if the letter arrived and to arrange an interview next week. The callee, the director of personnel, is eager to arrange an interview. The class will assess the skills of both the caller and the callee.
3. Role play a group discussion among five or six classmates. The group is to reach a consensus on one of the following situations:
   - **a.** A donor has offered $5000 for a scholarship in a professional field. The group is to decide how to implement the scholarship.
   - **b.** A department in a business has $5000 budgeted for employee travel to professional conferences and seminars around the country for the fiscal year. There are ten members in the department; each member has two or three conferences he or she would like to attend. Expenses would range from $200 for short, in-county seminars to $2000 for week-long, out-of-state conferences. The group must decide on the criteria to spend the money. Following the discussion the class will assess the verbal and nonverbal skills of the participants.
4. Prepare a 6- to 10-minute extemporaneous oral report with visual materials for one of the following situations:
   - **a.** *A persuasive speech.* Choose a controversial subject (four-day work week, class registration procedures, bookstore policies, degree requirements in your field, and so on). Prior to your speech, pass out index cards to the class and ask each member to write opinions on this subject. Deliver your speech. Each classmate will evaluate your presentation by rating you according to the rating blank in Figure 13.3. Following the speech, ask your classmates to read their opinion cards and discuss how you have or have not changed their opinions.
   - **b.** *An informative speech.* Refer to the chapters on instructions, descriptions of mechanisms, and analysis of a process. Select one of the report strategies; prepare and deliver an informative oral report. Each classmate will assess your presentation by filling out the appropriate rating blank in Figures 13.4, 13.5, or 13.6.

APPENDIX A

CONVENTIONS OF CONSTRUCTION, GRAMMAR, AND USAGE

INTRODUCTION

Effective communication depends not only on content and format but also on precise adherence to the conventions of sentence construction, grammar, and usage. This section will provide you with a brief handbook, exercises, and reference tables of professional and technical writing conventions.

The conventions that govern written English are not prescriptive rules. Rather, they are patterns developed by careful writers over decades and accepted by publishers, professional and technical writers, educators, and the public. Because English is a living language, these conventions alter over periods of time. For instance, it has long been conventional to use a comma before *and* in a written series of three or more items (i.e.: *He assembled the nuts, bolts, and rivets*). In recent years some publishers, businesses, and industries have agreed to omit the comma (i.e.: *He assembled the nuts, bolts and rivets*). Those who are concerned with language classify the conventions as *standard English* and *general English*. The majority of professional and technical writers adopt standard English conventions; therefore, this appendix covers the more formal conventions. Use these conventions to edit your texts and to troubleshoot your writing problems.

THE SENTENCE

Main Sentence Elements

A sentence is a group of words containing a subject and verb and expressing a complete thought. English sentences are arranged in many word order patterns which convey meaning. Lewis Carroll's nonsense sentence

> T'was brilig and the slithy toves did gyre and gimble in the wabe.

reveals patterns which make sense to English-speaking people. We recognize a subject and verb (*T'was*), the conjunctions (*and*), a helping verb (*did*), a preposition (*in*). Using the pattern we can develop any number of sentences:

> It was Monday, and the electronic technicians did assemble and test in the plant.
>
> It was noon, and the construction workers did toil and sweat in the sun.

There are five basic sentence patterns in English. Subjects, objects, and complements are nouns or pronouns. Verbs are words which express action (*jump*) or state of being (*is, seem*).

Pattern 1: Subject + Verb

S + V
Water boils.

Pattern 2: Subject + Verb + Direct Object

S + V + DO
Foremen supervise construction.

Pattern 3: Subject + Linking Verb + Subjective Complement

S + V + SC
Helium is a gas.

(A subjective complement renames the subject.)

Pattern 4: Subject + Verb + Indirect Object + Direct Object

S + V + IO + DO
The company gave Ms. Barnes a plaque.

(An indirect object is the receiver of the direct object.)

Pattern 5: Subject + Verb + Direct Object + Objective Complement

S + V + DO + OC
The union elected him president.

(An objective complement renames the direct object.)

There are actually many patterns beyond these basic five and many ways to invert or expand a sentence with single, one-word adjectives and adverbs or multi-word phrases and clauses. Consider that a Pattern 2 sentence

S + V + DO
The man has a compass.

may be inverted to ask a question

Has the man a compass?

or expanded to

adjective clause
The man who shares my drafting cubicle has

adjective **prepositional phrase**
a metric compass in his hand.

Recognition of the basic elements allows you to construct sentences which make sense and to punctuate the elements according to the conventions.

Secondary Sentence Elements

Secondary sentence elements are typically used as modifiers; that is, they describe, limit, or make more exact the meaning of the main elements.

Adjectives and adverbs. Single words used as modifiers are related to the words they modify by word order. Adjectives modify nouns or pronouns and usually stand before the word modified.

ADJ ADJ
He is a precise, concise writer. (modifies writer)

Adjectives may stand after the modified noun or pronoun or follow a linking verb:

N ADJ
The temperature made the metal brittle. (modifies metal)

N ADJ
The metal was brittle. (modifies metal)

Adverbs, which modify verbs, adjectives, or other adverbs, are more varied in position. They usually stand close to the particular word or element modified.

ADV
He worked late. (modifies the verb worked)

ADV
He worked quite late. (modifies the adverb late)

ADV
He was rather late. (modifies the adjective late)

ADV
He had never been late. (modifies verb phrase had been)

ADV
Unfortunately, he was always late. (modifies whole sentence)

Edit your writing so that the modifiers are clearly related to the words or statements they modify.

Ambiguous:	The plasterers have finished the wall almost. (Almost seems to modify wall.
Clear:	The plasterers have almost finished the wall.
Ambiguous:	I only need a few seconds.
Clear:	I need only a few seconds.
Ambiguous:	The red brick doctor's office.
Clear:	The doctor's red, brick office.

Do not place an adverb within an infinitive phrase (to read, to plan, to construct).

Split infinitive:	We have to further plan the assembly.
Clear:	We have to plan further the assembly.

Exercise

Add adjectives or adverbs within each sentence. If more than one position is possible, explain what change of emphasis would result from shifting the modifier.

1. **Add only.** Last week I ordered a computer package with a spelling checker.
2. **Add definitely.** I am convinced that you are the one to complete the audit.
3. **Add hardly.** Although I adjusted the lever, I could hear the bass resonate.
4. **Add engineer's.** The well-worn report lay on the shelf.
5. **Add carefully.** The committee has to read all of the reports.

Phrases as modifiers. A phrase is a group of related words without a subject or verb. It cannot stand alone. A phrase is connected to a sentence or to one of its elements by a preposition or a verbal.

Prepositional phrases. A prepositional phrase consists of a preposition (*in, at, by, from, under, at,* etc.) followed by a noun or pronoun plus, possibly, modifiers. It functions as an adjective or adverb, depending on what element it modifies.

> He entered from the door (modifies the verb entered) of my office (modifies the noun door).

Verbal phrases. A verbal phrase consists of a participle, gerund, or infinitive (verb forms without full verb function) plus its object or complement and modifiers. A participial phrase functions as an adjective; a gerund phrase as a noun; and an infinitive phrase as either a noun, adjective, or adverb.

Participial phrase: Circuit boards containing any defects should be scrapped. (modifies boards)

Gerund phrase: Drafting the blueprint was the next task. (functions as noun subject)

Infinitive phrase: To conduct a market survey (functions as noun subject) is the easiest way to determine the cost. (functions as adjective modifying way.)

Place modifying phrases next to the word modified. Participial and infinitive phrases will give you the most trouble.

Misrelated: He distributed notebooks to the trainees bound in plastic. (The participial phrase seems to modify trainees.)

Revised: He distributed notebooks bound in plastic to the trainees.

Misrelated: The man who was lecturing to emphasize a point pounded the podium. (The infinitive phrase seems to modify lecturing.)

Revised: The man who was lecturing pounded the podium to emphasize a point.

Dangling Participial Phrase: Looking up from my desk, Jane gave me the report. (The phrase seems to modify Jane.)

Revised: Looking up from my desk, I accepted the report from Jane.

Clauses. A clause is a group of words that contains a subject and verb plus modifiers. An **independent** (main) **clause** is a complete expression which could stand alone as a sentence. A **dependent** (subordinate) **clause** also has a subject and verb but functions as part of a sentence. It is related to the independent clause by a connecting word which shows its subordinate relationship either by a relative pronoun (*who, which, that,* etc.) or a subordinate conjunction (*because, although, since, if,* etc.).

Independent clauses: The water tastes brackish because it is contaminated.

The laser beam penetrated the metal plate, and the plate glowed red.

Dependent clauses:

If your engine is hot, add antifreeze.

The drive belt which slipped shredded.

After I attached the heat sink, the rectifier cooled.

We will make a final test because a dry run was never completed.

Sentence Classification

Sentences may be classified according to the kind and number of clauses they contain as simple, compound, complex, or compound-complex.

Simple sentences. A simple sentence contains an independent clause and no dependent clauses. It may contain any number of modifiers or compound elements.

> The capsule exploded.
>
> The tiny, white, plastic capsule expanded and exploded due to the high temperature in the storage bin.

Compound sentences. Compound sentences contain two or more independent clauses and no subordinate clauses. They may be joined by coordinating conjunctions (*and, or, but,* etc.), semicolons, or conjunctive adverbs (*nevertheless, therefore, however,* etc.).

> A dot matrix printer is acceptable, but a daisy wheel printer produces easier-to-read copy.
>
> A Radio Shack computer is flexible; it allows you to print hard copy of graphics.
>
> A personal computer is expensive; nevertheless, it is a practical tool for the professional writer.

Complex sentences. A complex sentence contains one independent clause and one or more dependent clauses.

> Because it is raining, the slump test will be postponed.
>
> The engineer who originally specified seven pilings changed her mind when she considered the sand content of the soil.

Compound-complex sentences. A compound-complex sentence contains two or more independent clauses and one or more dependent clauses.

> Because the text is illustrated with tables and sample materials, it is an indispensable guide for technical writers, and it may be used by students and professional writers in business and industry.

Complex sentences are used more frequently than simple, compound, or compound-complex sentences in published writing today. Complex sentences allow for more variety than simple sentences and allow the writer to manipulate emphasis. Recognition of independent and dependent clauses is necessary also for adding conventional punctuation.

Exercise

1. Combine these simple sentences as directed.
   - **a.** Combine into a compound sentence.
     A v-t voltmeter measures sine waves. An oscilloscope measures nonsinosuidal voltage.
   - **b.** Combine into a complex sentence.
     Pressure-sensitive tapes serve the electrical industry well. They insulate all manner of equipment and last longer than other tapes.
   - **c.** Combine into a compound-complex sentence.
     Magnetic memories can store more digital information on a par with optical disks. Optical disks can recreate visual images at a lower cost than magnetic memories. Optical disks can recreate visual images at a higher speed.
2. Rewrite these complex sentences to shift the emphasis as directed.
   - **a.** Emphasize the idea that he disliked his car rather than the idea that it uses too much gas.
     He disliked the car because, as far as I could determine, it used too much gas.
   - **b.** Emphasize the idea that soaking will separate the components rather than the idea that you can separate them in acids.
     If you like, you can separate the components by soaking them in acids.

BASIC SENTENCE ERRORS

Sentence Fragments

If a group of words is written as a sentence, but the group lacks a subject or a verb or cannot stand alone independently, it is called a sentence fragment. A fragment may be corrected by adding a subject or verb, joining it to another sentence, or rewriting the passage in which it occurs.

Fragment: The siphons, which were described earlier.

Revision: The siphons, which were described earlier, must be ordered. (verb added)

Fragment: The company continues to lose money. Although production has increased.

Revision: The company continues to lose money although production has increased. (joined to another sentence)

Fragment: Effective writing requires many skills. For example, a command of conventional grammar and the application of correct mechanics.

Revision: Effective writing requires many skills, such as a command of conventional grammar and the application of correct mechanics. (rewritten and combined into a sentence)

Exercise

Rewrite or combine these fragments into complete sentences.

1. Although we are trying to please the personnel of each department by presenting a variety of training programs.
2. The late arrivals having been named in this report along with the reasons for their tardiness.
3. The arrangements should be checked immediately. Registrants to be confirmed immediately.
4. Each month more than $1500 can be saved if the department buys a copy machine. No decrease in quality if the machine is top quality.
5. I recommend we use copper. Not zinc.

Sentence Parallelism

Parallel structure involves writing related ideas in the same grammatical constructions. Adjectives should be parallel with other adjectives, verbs with verbs, phrases with phrases, and clauses with clauses.

Nonparallel: Tungsten steel alloys are tough, ductile, and have strength. (adjective, adjective, verb plus noun)

Parallel: Tungsten steel alloys are tough, ductile, and strong. (all adjectives)

Nonparallel: He must learn the language and to be knowledgeable about his computer. (verb and infinitive)

Parallel:	He must learn the language and become knowledgeable about his computer. (both verbs)
Nonparallel:	He will be hired if he has the required training, if he has three years experience, and by being recommended by his former employer. (dependent clause, dependent clause, phrase)
Parallel:	He will be hired if he has the required training, if he has three years of experience, and if he is recommended by his former employer. (all dependent clauses)

Edit your writing to ensure that parallel ideas are expressed in parallel structures.

Exercise

Rewrite the nonparallel structures.

1. The computer is inexpensive, compact, and it is easy to use.
2. Before studying architecture, you should assess whether you have design ability and if you are exacting with numbers.
3. To write well one must be able to organize materials, have a flexible vocabulary, and one should know grammatical and mechanical conventions.
4. He was well liked and had training in management skills.
5. She attached the sphygmomanometer by positioning the patient, placing the cup above the elbow, and the clasp was secured.

Run-on and Comma-spliced Sentences

A run-on (also called a fused) sentence occurs when two or more independent clauses are written as one sentence without appropriate punctuation. A run-on sentence may be corrected by separating the fused clauses with a period or semicolon or by rewriting the sentence.

Run-on:	Ace Company will revise its maternity policy men will be eligible for child-rearing leave.
Correction:	Ace Company will revise its maternity policy; men will be eligible for child-rearing leave.

or

Ace Company will revise its maternity policy by allowing men to be eligible for child-rearing leave.

A comma-spliced sentence occurs when two or more independent clauses not joined by a coordinating conjunction or conjunctive adverb are written with only a comma between them.

Splice: Employees are entitled to eight sick days per year, they may be concurrent.

Correction: Employees are entitled to eight sick days per year; they may be concurrent.

or

Employees are entitled to eight sick days per year. They may be concurrent.

Splice: The restaurant requires a deposit for our annual dinner engagement, therefore, we must send a $50.00 check.

Correction: The restaurant requires a deposit for our annual dinner engagement; therefore, we must send a $50.00 check.

Using your understanding of clause structure, edit your writing to avoid run-on and comma-spliced sentences.

Exercise

Correct these run-on and comma-spliced sentences.

1. This guide is a product of months of research, compilation was done by Specialist Jim Smith.
2. Adult education opportunities are plentiful, moreover, all classes may be offered at our training facility.
3. After three straight days of bargaining, the talks broke down they will resume on Monday.
4. The union refused to consider benefit reductions but it did express willingness to negotiate increased work hours.
5. Mr. Larsen has participated in other civic activities in addition to his involvement in public schools he is a member and past officer of the Chamber of Commerce.

AGREEMENT

The most common grammar errors are subject-verb disagreements and pronoun-antecedent disagreements. For instance, if a subject is plural, then its verb must be plural as well, and if a pronoun antecedent is singular, then its pronoun must be singular as well.

Subject and Verb Agreement

Verbs change form from the singular to the plural. Consider:

Singular	***Plural***
I am	We are
He is	They are
She was	They were
He edits	They edit

A verb must agree with its subject in number. Generally, English-speaking people make these alterations automatically, but problems arise in a variety of structures.

In a sentence with compound subjects joined by *and* the verb is plural unless the subjects are considered a unit.

> The technician and engineer **are** consulting. (two persons)
>
> The accountant and auditor **reviews** my books monthly. (one person)

In a sentence with compound subjects joined by *or, nor, either . . . or, neither . . . nor* the verb usually agrees with the closest subject.

> The diodes or the transistor **is** faulty.
>
> Neither the transistor nor the thermistors **are** operating.

A singular subject followed by a phrase introduced by *as well as, together with, along with, in addition to* ordinarily takes a singular verb.

> The president as well as the vice-president **was held** responsible for the mismanagement.

Collective nouns (*committee, jury, crowd, team, herd,* etc.) usually take a singular verb.

> The committee **is meeting** Tuesday.
>
> The jury **is arguing** with the judge.

When a collective noun refers to members of the group individually, a plural verb is used.

> The jury **are arguing** among themselves.

Expressions signifying quantity or extent (*miles, years, quarts,* etc.) take singular verbs when the amount is considered as a unit.

> Ten dollars **is** too much to pay for a tablet.
>
> Six hours **is** too long to work without lunch.

A singular subject followed by a phrase or clause containing plural nouns is still singular.

> The highest number of diesel trucks **is** produced in Europe.
>
> The nurse who tends the heart patients **finds** them to be grateful.

When a sentence begins with *there is* or *there are,* the verb is determined by the subject which follows.

> There **are** an estimated 100 employees in this building.
>
> There **is** a conflicting opinion over capital punishment.

A verb agrees with its subject and not with its complement.

> Our chief trouble **was** (not were) malfunctions in the testing equipment.

Exercise

Select the appropriate verb to agree with its subject.

1. Three semesters (*is, are*) not enough to master French.
2. The bulk of our tax dollars (*go, goes*) to defense spending.
3. He is one of those people who (*is, are*) always willing to help.
4. The best benefit (*is, are*) the vacations.
5. Either the man or the woman (*assist, assists*) me with the payroll.

Pronoun Agreement

A pronoun is a word which takes the place of a noun, such as *he, who, itself, their, ourselves,* etc. (A complete list of pronouns is reviewed in Table A.1.)

> The technician checked **his** circuit boards.

The woman **who** trained me could assemble the parts **herself**.

A pronoun must agree in number with the word for which it stands, its antecedent.

Faulty: Ace Company is furloughing twenty of **their** employees.
Correct: Ace Company is furloughing twenty of **its** employees.

Faulty: Send the receipts to the bookkeeping department. **They** will issue the refunds.

Correct: Send the receipts to the bookkeeping department. **It** will issue the refunds.

Faulty: Anyone can take **their** accrued sick leave when necessary.
Correct: Anyone can take **his** (or **his or her**) accrued sick leave when necessary.

Table A-1 Pronouns

PERSONAL PRONOUNS:			
	Subject	Object	Possessive
First person			
Singular	I	me	my, mine
Plural	we	us	our, ours
Second person			
Singular & plural	you	you	your, yours
Third person			
Singular			
masculine	he	him	his
feminine	she	her	her, hers
neuter	it	it	its
Plural	they	them	their, theirs
RELATIVE PRONOUNS:	who	whom	whose
	that	that	
	which	which	whose, of which
INTERROGATIVE PRONOUNS:	who	whom	whose
	which	which	whose, of which
	what	what	

REFLECTIVE AND INTENSIVE PRONOUNS:

myself, yourself, himself, herself, itself, oneself, ourselves, yourselves, themselves

DEMONSTRATIVE PRONOUNS: this, these, that, those

INDEFINITE PRONOUNS:

all	both	everything	nobody	several
another	each	few	none	some
any	each one	many	no one	somebody
anybody	either	most	nothing	someone
anyone	everybody	much	one	something
anything	everyone	neither	other	such

RECIPROCAL PRONOUNS: each other, one another

NUMERAL PRONOUNS: one, two, three . . . first, second, third . . .

When a pronoun's antecedent is a collective noun, the pronoun may be either singular or plural, depending on the meaning of the noun.

The committee planned **its** next meeting. (the unit)

The committee gave **their** reports. (the individual members)

Usually a singular pronoun is used to refer to nouns joined by *or* or *nor*.

Neither Jane nor Judy did **her** share.

When the antecedent is a common-gender noun (*customer, manager, instructor, student, supervisor, employee,* etc.), the traditional practice has been to use *he* and *his* as in

A manager routinely evaluates **his** employees.

However, writers who are sensitive to sexist elements of our language are more prone to use both *his* or *her* if the gender of the antecedent is not known.

A manager routinely evaluates **his or her** employees.

If *his or her* must be repeated frequently, the cumbersome usage may be avoided by changing the singular antecedent to a plural construction.

Managers routinely evaluate **their** employees.

Some indefinite pronouns (*some, all, none, any,* etc.) used as antecedents require singular or plural pronouns, depending on the meaning of the statement.

Everyone, everybody, anyone, anybody, someone, no one, and *nobody* are always singular.

> Everyone must turn in **his or her** timesheet.
>
> Somebody erased **his or her** floppy disk.

All, any, some, or *most* are either singular or plural, depending on the meaning of the statement.

> All of the employees received **their** payroll deduction forms. (All refers to employees and is plural; all is the antecedent of their.)
>
> All of the manuscript has been typed, but **it** has not been proofread. (All refers to manuscript and is singular; all is the antecedent of it.)

In standard English usage *none* is usually singular unless the meaning is clearly plural.

Standard:	None of the men finished **his** work.
General:	None of the men finished **their** work.
Clearly plural:	None of the new computers **are** as large as their predecessors. (The sentence clearly refers to all new computers.)

When a pronoun is used, it must have a clearly identified antecedent.

Ambiguous:	The CRT fell on the keyboard and broke **it.** (It could refer to CRT or keyboard.)
Clear:	The CRT broke when **it** fell on the keyboard.
Ambiguous:	The consultants recommended a new method of shipping parts. This is the company's best alternative for the future. (It is not clear if this refers to method or to the implied word recommendation.)
Clear:	The consultants recommended a new method of shipping parts. This recommendation is the best alternative for the future.

Exercise

Correct these misused pronouns.

1. The company had high hopes for the new research program, but they encountered financial problems.

2. Neither the foreman nor the laborers want his pay reduced.
3. Everybody supported their union.
4. An instructor should encourage his students to ask questions.
5. Electrical engineering is an interesting field, and that is what I want to be.

Pronoun Case

Review the pronoun table. The personal pronouns and the relative or interrogative pronoun *who* have three forms depending on whether the pronoun is used as a subject, an object, or a possessive. Writers frequently encounter a few problems in proper case usage.

The object form of a pronoun is used after a preposition.

Incorrect: The work was divided between **he** and **I**.
Correct: The work was divided between **him** and **me**.

Incorrect: That is the data processor about **who** I have spoken.
Correct: That is the data processor about **whom** I have spoken.

In written English *than* is considered a conjunction, not a preposition, and it is followed by the form of the pronoun that would be used in a complete clause, whether or not the verb appears in the construction.

I am more experienced than **she** [is].

I like him better than [I like] **her**.

In general usage many educated people say "It is me" or "This is her," but standard English usage prefers the subject form after the linking verb *be*.

It is **I**.

This is **she**.

That is **he**.

It will be **I** who fail.

Although the distinction between *who* and *whom* is disappearing in oral communications, standard English usage prefers the distinction in writing. *Who* is the standard form when it is the subject of a verb; *whom* is the standard form when it is the object of a preposition or the direct object.

That is the professor **who** taught me chemistry. (Who is the subject of the verb taught.)

That is the woman **whom** I recommended for promotion. (*Whom* is the direct object of *recommended.*)

To **whom** are you speaking? (*Whom* is the object of the preposition *to.*)

Reflexive and Intensive Pronouns

The reflexive form of a personal pronoun is used to refer back to the subject in an expression where the doer and recipient of an act are the same.

He had only **himself** to blame.

I timed **myself** typing.

The same form is sometimes used as an intensive to make another word more emphatic.

The raise was announced by the president **himself**.

Safety **itself** is crucial.

In certain constructions writers mistakenly consider *myself* to be more polite than *I* or *me*, but in standard English the reflexive forms are not used as substitutes for *I* or *me*.

Faulty: Mr. Jones and **myself** attended the meeting.
Correct: Mr. Jones and **I** attended the meeting.

Faulty: The work was completed by Ms Burns and **myself**.
Correct: The work was completed by Ms Burns and **me**.

Exercise

Select the correct pronoun in each of the following sentences.

1. From (*who, whom*) will we receive the instructions?
2. The Director of Training assigned the project to Jones and (*I, me*).
3. It is (*we, us*) who were to leave early.
4. She was later than (*I, me*).
5. Smith, White and (*I, myself*) were assigned to conduct the survey.

USAGE GLOSSARY

Many words in the English language are so similar that they cause confusion. Following is a brief glossary and exercise of such terms for you to review. (Words that are asterisked are corruptions of correct usage; they are not to be used in formal writing.)

accept, except

Accept means "receive" or "agree to."

Community colleges accept a wide variety of students.

As a preposition, *except* means "other than."

I did all of the work except your report.

As a verb, *except* means "exclude," "omit," "leave out."

If you except Mr. Jones, no other president has owned his own Lear Jet.

advice, advise

Advice is a noun meaning "guidance."

If I wanted it, I would ask for your advice.

Advise is a verb meaning "counsel," "give advice to," "recommend," or "notify."

I advise you to exercise stock options.

affect, effect

Affect means "change," "disturb," or "influence."

The rising cost of fuel has drastically affected the trucking industry.

It can also mean "feign" or "pretend to feel."

Although she knew she was to be promoted, she affected surprise when notified.

As a verb, *effect* means "bring about," "accomplish," or "perform."

She effected a perfect word-processed report.

As a noun, *effect* means "result" or "impact."

Her extra work had no effect on the vice-president.

all ready, already

Use *all ready* when *all* refers to things or people.

At noon the secretaries were all ready to lunch.

Use *already* to mean "by this time" or "by that time."

I have already typed that report.

all right, *alright

All right means "completely correct," "safe and sound," or "satisfactory."

My answers to the interviewer were all right. (The meaning is that *all* of the answers were *right.*)
Despite a few cuts, I was all right.

Do not use *all right* to mean "satisfactorily" or "well."

*Do not use *alright* anywhere; it is a misspelling of *all right.*

almost, *most all

Almost means "nearly."

By the time we reached Miami, the tank was almost empty.

*In formal writing, do not use *most all*; use *almost all* or *most.*

Almost all (or most) of the employees were eligible for vacation.

a lot, *alot

A lot of and *lots of* are colloquial and wordy. Use *much* or *many.*
*Do not use *alot* anywhere; it is a misspelling of *a lot.*

as, as if, like

Use *as* or *as if* to introduce a clause.

As I drove into the parking lot, my tire blew out.
He looked as if he had worked all night.

Use *like* to mean "similar to."

The logo looked like ours.

assure, ensure, insure

Assure means "state with confidence to."

I assure you that he will be hired.

Ensure means "make sure" or "guarantee."

There is no way to ensure that every policy is understood.

Insure means "make a contract for payment in the event of specified loss, injury, or death."

He insured the package for $100.00.

complement, compliment

As a verb, *complement* means "bring to perfect completion."

His red tie complemented his blue suit.

As a noun *complement* means "something that makes a whole when combined with something else" or "the total number of persons needed."

Practice is the complement of learning.

Without a full complement of workers, we cannot complete the task.

As a noun, *compliment* means "expression of praise."

He complimented the appearance of my report.

continual, continuous

Continual means "going on with occasional slight interruption."

At the office there is a continual humming of typewriters.

Continuous means "going on with no interruptions."

We have 24-hour guard service; surveillance is continuous.

different from, *different than

*Do no use *than* after different. Use *from*.

His management style is different from mine.

few, fewer, little, less

Use *few* or *fewer* with countable nouns.

We have fewer employees than we did a year ago.

Use *little* or *less* with uncountable nouns.

I have less experience than you on the word processor.

good, well

Use *good* as an adjective, but not as an adverb.

The proposal for staggered work hours sounded good to many employees.

Use *well* as an adverb when you mean "in an effective manner," "ably."

He did so well on the project that he was promoted.

Use *well* as an adjective when you mean "in good health."

She hasn't looked well since her operation.

***hopefully**

Use *hopefully* to modify a verb.

She looked hopefully at her boss as he scanned her proposal.

*Do not use hopefully when you mean "I hope that," "we hope that," or the like.

Incorrect: Hopefully, the company will make a profit this quarter.

Revised: The stockholders hope that the company will make a profit this quarter.

its, it's

Its is the possessive form of *it.*

I like this company because of its location and its benefits.

It's means "it is."

It's evident that we are an expanding company.

stationary, stationery

Stationary means "not moving."

The typewriter was on a stationary table.

Stationery means "writing paper."

We had to order more stationery from the printing department.

Exercise

Supply the correct word. You may have to change the tense of the verbs.

1. Everyone has ______________ the invitation ____________ Sam. (accept, except)
2. I ______________ you to follow your instructor's ________________ . (advise, advice)
3. The malfunctioning air conditioning __________________ our tempers.
   The manager ___________________ a defiant look.
   The ____________________ of nuclear fallout are under study.
   (affect, effect)

4. Finally, the reports were xeroxed, and we were ________________ to begin the board meeting.
   The President had ______________ left for lunch when I reported for our interview.
   (already, all ready)
5. It is ________________ with me if you use correction tape.
   (alright, all right)
6. Winstons taste good ________________ a cigarette should.
   (as, as if, like)
7. I ________________ you that we can ________________ your right to strike.
   (ensure, assure, insure)
8. If there are ________________ members, it means ______________ work for the secretary.
   (fewer, less)
9. The departmental members work ________________ together, and I feel ________________ about their cooperation.
   (good, well)
10. Ace Company is an ideal employer because of ________________ benefits, and ________________ improving ________________ stock option program each year.
    (its, it's)

APPENDIX B

PUNCTUATION AND MECHANICAL CONVENTIONS

INTRODUCTION

The professional or technical writer must be conscious and demanding of punctuation and mechanical conventions to prevent vagueness and misreading. Two practices of punctuation are prevalent today. A few businesses and industries adopt an open punctuation system, which favors only essential marks and omits those that can be safely omitted, such as the comma before *and* in a series (screws, nuts, and bolts). The majority of professional writers use standard, or close, punctuation conventions because these conventions promote greater accuracy. This appendix reviews close punctuation conventions.

One of the major characteristics of professional and technical writing is the extensive use of abbreviations, numbers, symbols, and other mechanics. This appendix reviews the general conventions that govern mechanics.

Both the punctuation and mechanical conventions are arranged in alphabetical order to help you edit your writing quickly.

PUNCTUATION

1.0 Apostrophe

1.1 Use an apostrophe to indicate the possessive case of the noun:

The company's product

Jack and Bob's office (joint possession)

Bill's or Jack's car (individual possession)

his sister-in-law's law practice

1.2 Use an apostrophe to indicate the possessive case of indefinite pronouns:

another's tools	neither's wrench
anybody's desk	one's customers
each one's station	somebody's computer

Do *not* use an apostrophe to indicate the possessive case of personal pronouns:

his schedule
Ours is the newest model.
The mistake was hers.
Its handle is steel.

1.3 Use an apostrophe to indicate the omission of letters in contractions:

I'm	o'clock
he'll	we're
can't	you're

Do not confuse *they're* with *their* or *there.*

Their supervisor knows they're there.

Do not confuse *it's* (it is) with *its* (a possessive).

It's demonstrating its graphic function.

1.4 Use an apostrophe to indicate the plural of letters, numbers, symbols, and cited words:

Your *r*'s look like your *n*'s.

You use too many *and*'s.

Your *7*'s look like your *l*'s.

the 1980's

The &'s are broken on all of the typewriters.

2.0 Brackets

2.1 Use brackets within a quotation to add clarifying words that are not in the original:

Mr. Roberts stated, "They [computers] have revolutionized his business."

2.2 Use brackets within a quotation to enclose the Latin word *sic* ("so," "thus") which indicates that a misspelling, grammatical error, or wrong word was in the original:

He wrote, "Your [sic] selected to head the committee."

3.0 Colon

3.1 Use a colon after a formal salutation:

Dear Ms Benson:

Gentlemen:

Good morning:

3.2 Use a colon to introduce a phrase or clause which explains or reinforces a preceding sentence or clause:

Food processing consists of three main steps: selecting the blade, measuring the ingredients, and processing at the appropriate speed.

The position sounds attractive: the salary is high and the opportunities for advancement are excellent.

3.3 Use a colon when a clause contains an anticipatory expression (*the following, as follows, thus, these*) and directs attention to a series of explanatory words, phrases, or clauses:

The requirements for the position are as follows:

1. a master's degree,
2. three years experience, and
3. willingness to relocate.

3.4 Use a colon to express ratios, to separate hours and minutes, and to indicate other relationships.

3:1	signal:noise
A:B	8:25 P.M.
Acts: 14:7	12:101–104 (volume 12, pages 101–104)

3.5 Use a colon between the main title and subtitle of a book:

Technical Writing: An Easy Guide

4.0 Comma
Refer to Table B.1 for a review of comma and semicolon usage in compound and complex sentences.

4.1 Use a comma to separate independent clauses joined by a coordinating conjunction.

The cursor shows where you are typing, and it moves across the screen as you type.

4.2 Use a comma after an introductory dependent clause:

If you have a two-drive computer system, you place your program diskette in drive A.

4.3 Use a comma after a conjunctive adverb introducing a coordinate clause:

This system is easy to use; however, we suggest that you read the directions carefully.

Table B.1 Comma and Semicolon Review for Compound and Complex Sentence Construction

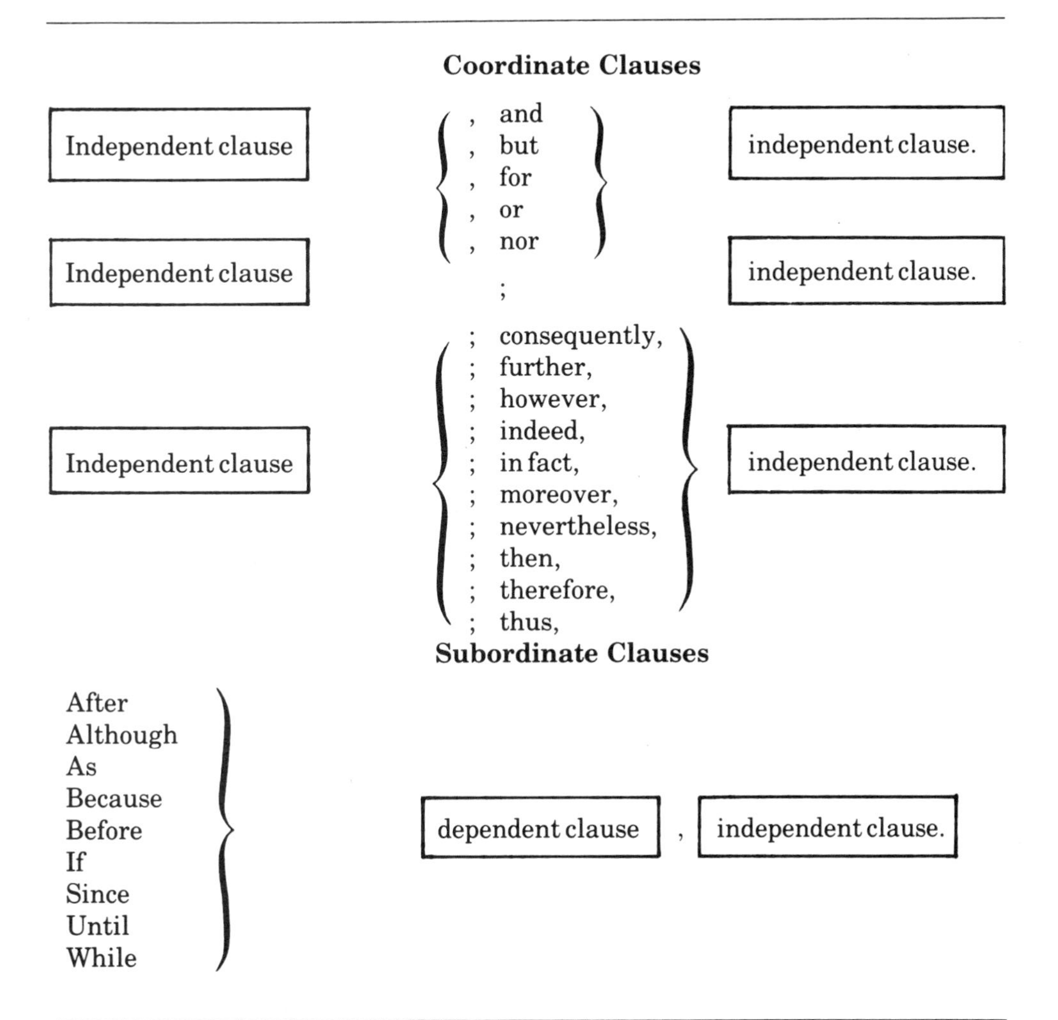

4.4 Use a comma to separate a nonrestrictive word, phrase, or clause from the rest of the sentence:

You should, however, make copies of your original diskette for safekeeping.

His goal, to become computer literate, is easy.

The cursor, which is a blinking square, shows where the entry will appear on the screen.

4.5 Use a comma to separate items in a series:

These instructions will teach you how to create, edit, or proof a file.

The Personnel Department will receive all letters of application, forward them to the appropriate search committee, and handle all correspondence.

The comma is often omitted in company names:

Jones, Smith and Scully

4.6 Use a comma to separate a series of adjectives or adverbs not connected by a conjunction:

Store your thin, sensitive, magnetic disks out of direct sunlight.

The computer blinked haphazardly, noisily.

4.7 Use a comma in a date to separate the day and year:

June 9, 1987

the April 15, 1985, deadline

Do *not* use a comma in the military or British form:

9 June 1987

4.8 Use a comma to separate city and state, state and county, county and state:

Fort Lauderdale, Broward Country, Florida, has twenty-seven electronic firms.

4.9 Use a comma to separate titles from names and to set off appositives:

Juan Murillo, M.D.

Julie Koenig, attorney-at-law

K. D. Marshal, Jr.

Robert H. Larsen, president

Jerry Noosinow, the PanAm pilot, checked the KAL log.

4.10 Use commas to group numbers into units of threes in separating thousands, millions, and so forth:

3,845
74,763
9,358,981

4.11 Use a comma after the salutation of an informal letter and after the complimentary close of most letters:

Dear Sally,

Very truly yours,

Cordially,

4.12 Use a comma to separate corporation abbreviations for company names:

Trolleys, Ltd.

Jones and Scully, Inc.

Ace Company, Inc.

4.13 Use a comma after expressions that introduce direct quotations:

President Dan Barker said, "We must double our profits."

If the phrase interrupts the quotation, it is set off by two commas:

"We must," said President Dan Barker, "double our profits."

5.0 Dash

5.1 Use dashes to set off emphatic and abrupt parenthetical expressions:

The idea of this program—it has been tested thoroughly—is to simplify spelling correction.

5.2 Use a dash to mark sharp turns in thought:

He was an arrogant man—with little to be arrogant about.

5.3 Use dashes to separate nonrestrictive material which contains commas from the rest of the sentence:

These manuals—*Guide to Operations, Disk Operating System, Word Proof,* and so on—are protected by copyright.

6.0 Ellipsis

6.1 Use the ellipsis (three spaced periods) to indicate any omission in quoted material:

Martin stressed, "The technical writer . . . must master punctuation." (The words *as well as the professional writer* have been omitted.)

6.2 Use four spaced periods to indicate the ellipsis at the end of a sentence.

The consultant stressed, "Write carefully . . . . " (The words *and edit endlessly* have been omitted.)

7.0 Exclamation Point

7.1 Use an exclamation point at the end of an exclamatory sentence to show emotion or force:

Shred the files!

We will not file bankruptcy!

7.2 Use an exclamation point at the end of an exclamatory phrase:

What a disaster!

8.0 Parentheses

8.1 Use parentheses to enclose an abruptly introduced qualification or definition within a sentence:

You may place your program diskette in drive A and your storage diskette (the one with your file on it) in drive B.

8.2 Use parentheses to enclose a cross-reference within a sentence:

The ellipsis (see Section 6.0) is used in direct quotations.

8.3 Use parentheses to enclose figures or letters to enumerate points:

> To use this program (a) insert your DOS diskette in drive A, (b) turn on your computer, (c) type in the date, and (d) press the Enter key.

9.0 Period

9.1 Use a period to signal the end of declarative or imperative sentences:

> Diskettes are sensitive to extremes of temperature.
>
> Do not try to clean diskettes.

9.2 Use a period with certain abbreviations (see pages 418-20 for exceptions):

Dr.	Jr.	Nov.
A.M.	Mr.	J. C. Lewis
P.M.	Ph.D.	etc.

9.3 Use a period before fractions expressed as decimals, between whole numbers and decimals, and between dollars and cents:

.10	$3.50
3.6	$0.92 (or 98¢ or 93 cents)

9.4 Use a period after number and letter symbols in an outline:

I.
- **A.**
- **B.**
  - **1.**
  - **2.**

10.0 Question Mark

10.1 Use a question mark at the end of an interrogative question:

> Do you own a personal computer?

Do *not* use one after an indirect question.

> He asked me if I owned a personal computer.

10.2 Use a question mark in parentheses to indicate there is a question about certainty or accuracy:

This is the best (?) computer.

11.0 Quotation Marks

11.1 Use quotation marks to set off direct speech and material quoted from other sources:

Dr. William Haskell writes, "Before 1960 the thing rarer than a marathoner was a health professional trained to care for one. Most doctors," he points out, "forbade post-cardiac patients to do anything more vigorous than walk to the refrigerator."

11.2 Use quotation marks to indicate nonstandard terms, ironic terms, and slang words:

This is a "gimmick."

His "problem" was his genius I.Q.

He "got his act together."

11.3 Use quotation marks to indicate titles of articles, essays, short stories, chapters, short poems, songs, television and radio programs, and speeches:

I read the article "Ten Years of Sports Medicine" in *Runner's World.*

Do *not* use quotation marks around quoted material which requires more than four lines in your paper. Display a long quote by indenting it ten spaces from your regular margins and omitting the quotation marks.
Commas and periods are always placed *inside* the closing quotation mark:

"Yes," Ms Gloss said, "we have a swine flu epidemic."

He said "electrons," but meant "electronics."

Semicolons and colons are placed *outside* the closing quotation mark:

He said, "The trapped air bubble will leave honeycombs"; honeycombs are sections of little indentations.

He said, "The trapped air bubble will leave honeycombs": little indentations.

Question marks, exclamation points, and dashes are placed inside *or* outside the final quotation mark, depending upon the situation:

He asked, "Is the oscillator connected to the mixer?"

Did he say, "The assembly is constructed of heavy-gauge stainless steel"?

12.0 Semicolon

Refer to Table B.1 for a review of comma and semicolon usage in compound and complex sentences.

12.1 Use a semicolon between coordinate clauses not connected by a conjunction:

The new system will use low-powered transmitters; it is called a cellular radio.

12.2 Use a semicolon before a conjunctive adverb including a coordinate clause:

Retort pouches are like cans; however, they do not dent.

12.3 Use a semicolon before a coordinating conjunction introducing a long or loosely related clause:

Niobium, which is used primarily as an alloy, is a metallic element that resists heat and corrosion and hardens without losing strength; and it is widely available in Canada and South America.

12.4 Use a semicolon in a series to separate elements containing commas:

J. D. Smyth, member of the board; Carol Winter, president; Glenn Morris, committee chairperson; and I attended the conference.

13.0 Virgule (Slash)

13.1 Use a virgule to indicate appropriate alternatives:

Define and/or use the words in sentences.

13.2 Use a virgule to represent *per* in abbreviations:

17 ft/sec   12 mi/hr

13.3 Use a virgule to separate divisions of a period of time:

the April/May report

the 1986/87 academic year

MECHANICAL CONVENTIONS

1.0 Abbreviations

1.1 Avoid the overuse of abbreviations. (See Table B.2 for a list of common technical abbreviations.)

1.2 Explain an abbreviation the first time you use it:

He has worked for the Department of Transportation (DOT) and the Office of Mental Health (OMH).

1.3 Omit most internal and terminal punctuation in abbreviations:

BTU	lb
psi	ft
DNA	rpm

1.4 If the abbreviation forms another word, use the internal and terminal punctuation:

in.	A.M.
gal.	No.

1.5 Use uppercase (capital) letters for acronyms and degree scales:

NASA (National Aeronautics and Space Administration)
VHF (very high frequency)
OEM (original equipment manufacturer)
C (Centigrade)
F (Fahrenheit)

1.6 Use lowercase (small) letters for abbreviations for units of measure:

Table B.2 Common Technical Abbreviations

Abbreviation	Meaning
ac	alternating current
amp	ampere
A	angstrom
az	azimuth
bbl	barrel
BTU	British Thermal Unit
C	Centigrade
Cal	calorie
cc	cubic centimeter
circ	circumference
cm	centimeter
cps	cycles per second
cu ft	cubic foot
db	decibel
dc	direct current
dm	decimeter
doz	dozen
dp	dewpoint
F	Fahrenheit
f	farad
fbm	foot board measure
fl oz	fluid ounce
FM	frequency modulation
fp	foot-pound
fpm	feet per minute
freq	frequency
ft	foot
g	gram
gal.	gallon
gpm	gallons per minute
gr	gram
hp	horsepower
hr	hour
in.	inch
iu	international unit
j	joule
ke	kinetic energy
kg	kilogram
km	kilometer
kw	kilowatt
kwh	kilowatt hour
l	liter
lat	latitude
lb	pound
lin	linear
long	longitude
log	logarithm
m	meter
max	maximum
mg	milligram
min	minute
ml	milliliter
mm	millimeter
mo	month
mph	miles per hour
No.	number
oct	octane
oz	ounce
psf	pounds per square foot
psi	pounds per square inch
qt	quart
r	roentgen
rpm	revolutions per minute
rps	revolutions per second
sec	second
sp gr	specific gravity
sq	square
t	ton
temp	temperature
tol	tolerance
ts	tensile strength
v	volt
va	volt ampere
w	watt
wk	week
wl	wavelength
yd	yard
yr	year

gph (gallons per hour)
cc (cubic centimeters)
rpm (revolutions per minute)
mph (miles per hour)
bps (bits per second)

1.7 Write plural abbreviations in the same form as the singular:

17 in
47 lb
5 hr
30 gph
10 cc

2.0 Capitalization

2.1 Use standard English conventions.

2.2 Begin all sentences with a capital letter.

2.3 Capitalize all proper nouns (proper names, titles which precede proper names, book and chapter titles, languages, days of the week, months, holidays, names of organizations and groups, races and nationalities, historical events, names of structures and vehicles, and so forth:

John Doe	Ace Construction Company
Professor Jane Doe	American Federation of Labor
Introduction to Nursing	Caucasian
French	Jewish
Monday	the Korean War
October	the Statue of Liberty
Labor Day	a Ford Mustang

2.4 Capitalize adjectives derived from proper nouns:

English	Corinthian
Elizabethan	Asian

2.5 Capitalize words like *street*, *avenue*, *corporation*, and *college* when they accompany a proper name:

Elm Street
Forty-second Avenue
Ace Company, Inc.
Yale University
Broward Community College

2.6 Capitalize *north*, *east*, *midwest*, *near east*, and so on when the word denotes a specific location:

the South
the Midwest
the Near East
101 Northwest First Street

2.7 Capitalize brand names:

Kleenex tissues	Xerox photocopies
Scotch tape	a Frigidaire
a Formica counter	the Astro-Turf field
a Polaroid camera	Sanforized

3.0 Hyphenation

3.1 Use a hyphen between some compound names for family relationships:

Hyphenated:	brother-in-law's company
One word:	my stepmother's portfolio
Two words:	my half brother was graduated

3.2 Use a hyphen in compound numbers from twenty-one to ninety-nine and in fractions:

thirty-seven cartons
forty-third year
four-fifths of the book
one-eighth inch

3.3 Use a hyphen after the prefixes *all-*, *ex-*, *self-*, and before the suffix *-elect:*

all-American
ex-president
self-contained
president-elect

3.4 Use a hyphen in some compound nouns:

kilowatt-hour
dyne-seven
foot-pound

3.5 Use a hyphen in compound adjectives when the latter precedes the word it modifies:

alternating-current motor
closed-circuit television
high-pressure system
easy-to-build model

3.6 Use a hyphen between a number and a unit of measure when they modify a noun:

6-inch ruler
12-volt charge
a 3-week-old prescription

4.0 Italics

4.1 Use italics (underline in handwritten or typed material) to indicate the names of books, magazines, newspapers, and other complete works published separately:

the book *Introduction to Nursing*
the magazine *Newsweek*
the movie *The Right Stuff*
Dante's *Divine Comedy*
Word Proof: A Manual

4.2 Use italics to indicate the names of ships and planes:

the H.M.S. *Ark Royal*

the U.S.S. *Independence*

4.3 Use italics to indicate words, symbols used as words, and foreign words which are not in general English usage:

The prizewinning orchids were *Alleraia* Ocean Spray, *Bloomara* Jim, and *Guantlettara* Noel.
coup d'etat
deus ex machina
The word *thrombosis* is derived from the Greek word *thrombos* which means "a clot" or "a clump."
Your *9*'s look like *7*'s.

Do not italicize foreign expressions which are established as part of the English language, such as:

a priori	bona fide	habeas corpus	pro tem
ad hoc	carte blanche	laissez faire	resume
ad infinitum	etc.	per annum	status quo
	ex officio	pro rate	

5.0 Numbers

5.1 Handle numbers consistently in any one report.

5.2 Write out single digit numbers from zero through nine when the number modifies a noun:

five disks	two printers
three word processors	nine keyboards

5.3 Use numerals for zero through nine when the number modifies a unit of measure, time, dates, pages, chapters, sections, percentages, money, proportions, tables, and figures:

2 inches	section 9
3-second delay	2 percent
5 gph	a 4% increase
9 years old	$50
2:40 A.M.	$0.05 or .05 cents
June 9, 1986	1:9
9 June 1986	4 to 2 odds
page 7	Figure 2
Chapter 1	Table 6

5.4 Use numerals for decimals and fractions:

0.6	1/4 or 0.25
3.341	7/16 in.
3/5 or 0.6	6½ lb

5.5 Use numerals for any number greater than nine:

10 psi	237 lb
97 employees	101,400 people

5.6 Write out numbers which are approximations:

a half cup of coffee
a quarter of a mile farther
a fifth of the energy
approximately three times as often

5.7 Place a hyphen after a number of a unit of measure when the unit modifies a noun:

7-inch handle
6-inch-diameter circle
10½-lb box
27-gal. capacity

5.8 When many numbers, both smaller than and greater than nine, are used in the same section of writing, use numerals:

Buy 4 sheets of 8-inch by 11½-inch paper, 15 sheets of 8-inch by 20-inch paper, and 7 manila envelopes.

Exception: If none of the numbers are greater than nine, write them all out:

The office contains eight desks, seven chairs, six file cabinets, and seven typewriters.

5.9 When one number appears immediately after another as a part of the same phrase, avoid confusion by writing out the shorter number:

nine 50-watt bulbs
1000 fifty-watt bulbs
thirteen 20-pound packages
twenty-two 2,500-component circuit boards

5.10 Place a comma in numbers in the thousands:

1,000
17,276
427,928

5.11 Write numbers in the millions in one of two ways:

2,700,000 or 2.7 million
16,000,000 or 16 million
$1,500,000 or $1.5 million
72,110,427

5.12 Write numbers in the billions, trillions, quadrillions, and so on in numerals:

2,700,000,000
47,337,426,104,900

5.13 Do not begin a sentence with a numeral:

> Fifteen inches of rain fell.
> *not*
> 15 inches of rain fell.

6.0 Symbols

6.1 Use symbols sparingly.

6.2 Define symbols in your text. (See Table B.3 for common technical symbols.)

Table B.3 Common Technical Symbols

Symbol	Word
%	percent
°	degree
&	and
′	feet
″	inches
$	dollar
¢	cents
@	at (12 at $2.00 each)
+	plus
−	minus
×	times
÷	divide
‖	greater than or derived from
=	equals
F	Fahrenheit
C	Centigrade
Rx	take (on prescriptions)
θ	the Sun, Sunday
£	pound
#	number
Hg	mercury (the element)
☿	Mercury (the planet)
X	snow
↑	gas
Ω	ohm
S	Silurian soil

7.0 Spelling

7.1 Use a dictionary when in doubt about the proper and preferred spelling of a word. (See Table B.4 for a list of frequently misspelled words.)

Table B.4 Frequently Misspelled Words

accidentally	comparative	heroes	prominent
achievement	competitive	humorous	propaganda
acquaintance	consensus	immediately	psychology
amateur	contemptible	indispensable	pursue
analysis	convenience	irrelevant	questionnaire
anonymous	courageous	irresistible	receive
anxiety	criticism	knowledge	rhythm
appreciate	definitely	laboratory	schedule
arctic	descent	leisure	scissors
athletics	desirable	lieutenant	secretary
auxiliary	despair	lighting	seize
awkward	disappear	loneliness	separate
bachelor	discipline	maneuver	sergeant
beggar	efficient	meant	siege
beginning	eighth	medieval	similar
believe	eligible	minimum	sophomore
benefited	equipped	mortgage	souvenir
bookkeeper	exaggerate	necessary	subtle
breath	exercise	ninth	succeed
bulletin	exhausted	noticeable	successful
bureau	existence	ocasionally	surprise
business	familiar	occurred	synonym
calendar	fascinating	omitted	thoroughly
campaign	fatigue	opportunity	tragedy
caricature	fiery	parallel	twelfth
catastrophe	foreign	parlaysis	unforgettable
cemetery	forty	pastime	unmistakable
colonel	government	possibility	vacuum
coming	guarantee	privilege	vengeance
committee	height	procedure	weird

INDEX